Agricultural Exports of India

- Lessons from East Asia

About the book...

Promoting competitiveness is one of the major objectives of a country's trade strategy. In this context, specialisation in exports has to be on the basis of comparative advantage considerations. Much of the research in this area has focused on trade in industrial goods. The study of comparative advantage in agricultural exports has received scant attention. Moreover, work in this area has been done in the context of developed countries whose primary exports form a low proportion of their total exports.

However, the weightage of primary products in the export basket of developing economies of South Asia is quite high, necessitating analysis of comparative advantage in agricultural exports. The present work is intended to fill the vacuum in this area of research. Indeed, the underlying objective of the whole exercise is to provide a thrust to India's agricultural exports.

About the author...

Mrs. Santosh Sachdev received her M.A (Economics) degree from the Delhi School of Economics, University of Delhi. She did her M.Phil and Ph.D from the School of International Studies, Jawahar Lal Nehru University, New Delhi. With teaching and research experience of over 30 years, Dr. Sachdev has published research papers in recognised journals, such as *Foreign Trade Review* and *Indian Economic Review.* Presently, Dr. Sachdev is a Reader in the Department of Economics, Kalindi College, University of Delhi.

Agricultural Exports of India

- Lessons from East Asia

Santosh Sachdev

NEW NC CENTURY

New Century Publications
34, Gujranwala Town Part-2
Delhi-110 033 INDIA

First Edition 2000

New Century Publications
Tel.: 7138192, 7247805
Fax: +91-11-7464774
Email : indiatax@del2.vsnl.net.in
Website : http://www.nexusindia.com/indtax

First Edition 2000

Reprint: 2024

ISBN: 81-7708-003-2

Published by New Century Publications and printed at D.K. Fine Art Press (P) Ltd., Delhi-110 052

PRINTED IN INDIA

Key to Numeration

1 lakh	=	0.1 million
10 lakh	=	1.0 million
1 crore	=	10.0 million
100 crore	=	1.0 billion

Exchange Rate

The unit of currency in India is Indian Rupee

As on 19.1.2000

1 US Dollar	=	43.56 Rupees
1 UK Pound	=	71.32 Rupees
1 Euro	=	44.03 Rupees

Fiscal Year

Fiscal year in India runs from April 1 to March 31 of the following year.

Contents

Preface xii
List of Tables xiv
List of Abbreviations xvii

Chapters

1. Introduction **1-22**
1.1 Performance of Exports: Past Studies
1.2 Methodology
1.2.1 Classification of Agricultural and Allied Products
1.2.1.1 Alternative Classifications
1.2.1.2 Classification Used in the Present Study
1.2.2 Competitiveness of Exports
1.2.2.1 Methodology Used in General
1.2.2.2 Justification for the Methodology Adopted in the Present Study
1.2.2.3 Methodology Adopted in the Present Study

2. Export Performance of South Asian Countries **23-51**
2.1 Introduction
2.1.1 Development Performance of the Three Sub-regions: An Overview
2.1.1.1 South Asia
2.1.1.2 The Newly Industrialising Countries (NIC-4)
2.1.1.3 The Association of South East Asian Nations (ASEAN-4)
2.2 Causal Factors for Differences in Growth Rates among the Three Sub-regions
2.2.1 Outward Oriented Trade Policies
2.3 Performance of Exports: South Asia
2.4 Reasons for Poor Performance of Exports: South Asia
2.5 Factors Supporting Outward-oriented Trade Policies

2.6 Conclusions

3. The Determinants of Competitiveness **52-81**

3.1 Introduction

3.2 Determinants of Competitiveness

3.2.1 Hypotheses

3.3 Sources of Comparative Advantage

3.3.1 Introduction

3.3.2 Ricardian Model

3.3.3 Heckscher-Ohilin Model (H-O)

3.3.3.1 Leontief Paradox

3.4 The Neo-Factor Proportions Approach

3.4.1 Human Capital or Skill Explanation of Trade

3.5 The New Technology Approach

3.5.1 Stage of Production

3.5.2 Technological Gap and Product Cycle

3.5.3 Product Cycle Model

3.5.4 Availability Hypotheses

3.5.5 Preference Similarity

3.6 New Theories of Trade among Industrial Countries

3.7 Theoretical Validity of the H-O Model

3.8 Conclusion

4. Resource Endowment Structure and Comparative Advantage **82-109**

4.1 Introduction

4.2 Empirical Verification of the H-O Model of Trade: South Asia

4.2.1 Methodology

4.2.2 Resource or Factor Endowment

4.2.3 Relative Factor Endowment

4.2.4 Factor Intensity

4.2.5 Interaction between Endowment and Intensity

4.3 Comparative Advantage: South Asia (Empirical Results)

4.3.1 Absolute Factor Endowment: South Asia

4.3.2 Relative Factor Endowment: South Asia

4.3.3 Factor Intensity of Agricultural and Labour Intensive Products
4.3.4 Comparative Advantage: Interaction between Endowment and Intensity
4.4 Dynamic Nature of Comparative Advantage: South Asia
4.5 Conclusion

5. Comparative Advantage and Pattern of Trade 110-129
5.1 Introduction
5.2 Revealed Comparative Advantage: A Measure of Comparative Advantage
5.2.1 Basic Concepts
5.2.1.1 Trade-cum-Production Index: RACGO and BOWEN Indices
5.2.1.2 Trade only Indices
5.2.2 The Commodity Composition of Exports
5.3 Empirical Results
5.3.1 Revealed Comparative Advantage (RCA)
5.3.2 Ideal Comparative Advantage and Revealed Comparative Advantage
5.3.3 Composition of Exports: NIC-4, ASEAN-4 and South Asia
5.4 Composition of Exports vis-a-vis Comparative Advantage
5.4.1 Intersectoral Comparisons
5.4.2 Intercountry Comparisons

5.5 Conclusion

6. Empirical Analysis of Agricultural Exports: India 130-182

6.1 Introduction
6.2 Importance of Agricultural Export Strategy for Developing Countries
6.3 Importance of Agricultural Exports: India
6.4 Performance of Agricultural Exports: India

6.4.1 India's Agricultural Exports and Comparative Advantage
6.4.2 India's Share in World Agricultural Exports
6.4.3 India's Agricultural Exports in Relation to Sectoral Output and Manufactured Exports
6.5 Reasons for Poor Performance of Agricultural Exports: India
6.5.1 Macro Economic Policies
6.5.1.1 Fundamentals
6.5.1.2 Selective Policies
6.5.2 Rate of Growth in Agriculture
6.5.3 Agricultural Wastes
6.5.4 Inefficiency in Resource Management
6.6 Policy Inferences for Agricultural Exports: India
6.6.1 Identification of Commodities
6.6.2 Measures of Create Export Surpluses
6.6.2.1 Yield Improvement
6.6.2.2 Development of Human Capital and Basic Infrastructure
6.6.2.3 Reduction in Wastes
6.6.2.4 Efficiency in Resource Management
6.7 Conclusion

7. Implications of Trade Policy Reforms on India's Agricultural Exports **183-207**
7.1 Introduction
7.2 Trade Policy: India
7.2.1 Trade Policy Till the Fifth Five Year Plan
7.2.2 Trade Policy During the Sixth Five Year Plan
7.2.3 Trade Policy: Seventh Five Year Plan
7.2.4 Economic Reforms: 1991-92
7.2.5 Export-Import Policy (EXIM Policy) for the Eighth Plan Period (1992-93 to 1996-97)
7.2.5.1 Objective of New Trade Policy
7.2.5.2 Trade Policy Measures
7.3 GATT Agreement on Agriculture (Uruguay Round)

7.3.1 Introduction
7.3.2 Agreement on Agriculture
7.3.2.1 Domestic Support Commitments
7.3.2.2 Market Access
7.3.2.3 Export Subsidy Commitments
7.4 Implications of Trade Policy Reforms on India's Agricultural Exports
7.4.1 Implications of GATT on India's Agricultural Exports
7.4.1.1 Domestic Support
7.4.1.2 Market Access
7.4.1.3 Export Subsidy
7.5 Conclusion

8. Conclusions **208-228**
8.1 Introduction
8.2 Hypothesis
8.3 Basic Thrust of the Study
8.4 Findings of the Study: Chapters 1 to 7
8.5 Conclusions and Suggestions

Bibliography and Data Sources **229-250**

Preface

Competitiveness is a dynamic phenomenon evolving as an integral component of the development process. As economies grow, they accumulate capital, develop skills and generate better technologies. Thus, they move up the ladder of comparative advantage from specialisation in primary products to products which are intensive in capital, skill, technology and knowledge.

The present study is based on the fundamental premise that India being a low income agricultural economy, enjoys comparative advantage in agricultural products and labour intensive manufactures. An increase in the share of agricultural exports in total exports would lead to better performance of her external sector.

This book offers a critical analysis of comparative advantage in agricultural exports of South Asia , especially of India. In view of the vastness of the subject and data limitations, the macro analysis (i.e. analysis of sectoral composition of exports and comparative advantage) deals with South Asian economies in general. However, the micro analysis (i.e. identification of agricultural products having comparative advantage) has been attempted with special reference to India.

The book is compiled into 8 chapters.

Chapter 1, as an introductory chapter, outlines the approach for analytical purposes and sets forth the objectives, scope and methodology of the study. Chapter 2 analyses the export performance of South Asian countries as compared to NIC-4 and ASEAN-4. Chapter 3 explains the model of export determinants in three parts, i.e. determinants of competitiveness, sources of comparative advantage and theoretical validity of Heckscher-Ohlin model of trade. Chapter 4 examines the proposition (by empirical verification of H-O model) that South Asian countries have comparative advantage in agricultural products and labour-intensive manufactures. Chapter 5 analyses the composition of exports of these countries, revealing inconsistency between composition of trade and comparative advantage. Chapter 6 deals with the importance of agricultural exports of India and suggests a suitable strategy to boost them. Chapter 7 is devoted to the implications of trade policy reforms on India's agricultural exports. Chapter 8 summarises findings of the study.

I wish to gratefully acknowledge the invaluable assistance received from the following scholars in the preparation of this book: Professor (Dr.) Indra Nath Mukherjee of the School of International Studies, Jawahar Lal Nehru University, New Delhi; Professor S. Tendulkar and Professor V.N. Pandit of the Delhi School of Economics, University of Delhi; and Professor S.K. Sachdev of the Institute of Management Technology, Ghaziabad.

Delhi
January 26, 2000

Santosh Sachdev

List of Tables

Table 2.1: Growth of Real GDP and GDP per capita (Compounded annual change) South Asia, ASEAN-4 and NIC-4

Table 2.2: Ratio of Exports to GDP, 1970, 1978, 1985 and 1986 (percent at current prices) South Asia, ASEAN-4 and NIC-4

Table 2.3: Percentage Share of Exports of Selected Countries in Total World Exports

Table 2.4: Structure of Merchandise Exports/Imports of South Asian Countries, 1965 and 1988

Table 2.5: Constant Market Share Analysis of Changes in Export by Countries, 1965-1980

Table 2.6: Ranks of each of the components contributing to export growth by countries and areas for the period 1965-67 to 1978-80 and 1972-80 decomposition analysis

Table 4.1: Structure of Merchandise Exports/Imports of South Asian Countries, 1965 and 1988

Table 4.2: Impact on CA by Interaction between Endowment and Intensity

Table 4.3: Absolute Resource Abundance Ratio: Land and Labour, South Asia

Table 4.4: Relative endowment of Capital, land and labour (skilled and unskilled) South Asia, NICs and ASEAN-4

Table 4.5: Estimates of Factor Intensities: India (1979-80)

Table 4.6: Interaction Between Factor Intensity and Factor Abundance of Different Product Groups: South Asian countries

Table 5.1: Composition of Trade Aggregates

Table 5.2: Revealed Comparative Advantage

Table 5.3: Composition of Exports: South Asia, NICs and ASEAN-4

Table 6.1: Share of India's Agricultural Exports to World Agricultural Exports

Table 6.2: India's International Trade in Agricultural Commodities

Table 6.3: Reasons for Poor Performance of Agricultural Exports in India

Table 6.4: Changes in Consumer Prices (Per Cent Per Annum)

Table 6.5: Overall Budget Surplus/Deficit of Central Government (Per Cent of GDP)

Table 6.6: External Debt to Export Ratio (Per Cent)

Table 6.7: Expenditure by Central Government on Basic Education and Infrastructure (Percentage of Total Government Expenditure)

Table 6.8: Average Annual Growth Rate of Agricultural Production by Countries (Per Cent)

Table 6.9: Commodities Having Comparative Advantage i.e. Commodities with RCA Index above Unity 1980-96

Table 6.10: Percentage of India's Exports in World Exports of Respective Commodity Groups

Table 6.11: Average Yield (Kg./HA) of Selected Agricultural Crops in India, World, Asian Countries

Table 6.12: Performance of Infrastructure Sector in India

List of Abbreviations

ADB	Asian Development Bank
ADLI	Agricultural Demand Led Industrialisation
AMS	Aggregate Measurement of Support
ASEAN	Association of South East Asian Nations
BOP	Balance of Payments
DE	Developing Economies
DME	Developed Market Economies
DRC	Domestic Resource Cost
EOU	Export-Oriented Units
EPZ	Export Promotion Zone
ESRF	Economic and Scientific Research Foundation
FAO	Food and Agriculture Organisation
GATT	General Agreement on Trade and Tariff
GDP	Gross Domestic Product
GFCF	Gross Fixed Capital Formation
GNP	Gross National Product
IMF	International Monetary Fund
LDC	Least Developed Countries
NIC	Newly Industrialising Countries
NWDB	National Wasteland Development Board
RCA	Revealed Comparative Advantage
SAARC	South Asian Association for Regional Co-operation
SITC	Standard International Trade Classification
UN	United Nations
UNCTAD	United Nations Conference on Trade & Development
UNIDO	United Nations Industrial Development Organisation
WB	World Bank

Agricultural Exports of India
- Lessons from East Asia

Chapter 1

Introduction

The economic growth of Asia Pacific Regions, which encompasses the Western Basin from Korea in the north to Indonesia in the South, has been most impressive since 1970s. Asian growth however, has not been uniform. During 1970s and 1980s some of the countries emerged as newly industrialising countries, some were in the process of entering this group while others lagged behind. It will not be wrong to classify the resulting groups into three main categories:

(a) The high income, newly industrialising countries (NIC-4) namely HongKong, Singapore, Korea and Taiwan.

(b) The middle income emerging industrialising countries namely Indonesia, Malaysia, the Philippines and Thailand. These countries are commonly known as ASEAN-4 i.e. the four larger members of the Association of South East Asian Nations.

(c) The low income countries of South Asia namely Bangladesh, India, Pakistan, Nepal and Sri Lanka.

The development performance of three subregions of Asia depicts that countries at a similar stage of development in 1950s showed varied economic performance during the last three decades. The economic performance of NIC-4 since 1960s has been spectacular, surpassing that of any other group of countries in modern times. Almost the same can be said of ASEAN-4 (except the Philippines). But the economic performance of South Asian countries and the Philippines has been less impressive. If we draw on the wider evidence of the debate about the causes of development, voluminous literature on development suggests that though the pace of industrialisation and trade depends in part on initial conditions, factors and resource endowment, country size, location and social mores and on international environment, policy formulation and implementation by the governments of different countries is the principal cause of differentials in growth rate. In particular, trade policies of developing countries can affect trade as well as economic performance significantly. Trade has proved to be more effective instrument for promoting efficient structural change than autarky.

The rapid industrial growth of many of the Asian developing countries compared to other developing regions reflects the different types of trade policies they have adopted. Countries applying outward oriented development strategies had a better performance in terms of exports, economic growth and employment whereas, countries with inward orientation encouraged increasing economic difficulties.

Most of the Asian developing countries followed import substitution strategies, at least initially.

The NIC-4 shifted to export orientation by early 1960s. The ASEAN-4 began to liberalise trade in early 1970s. However, a pattern of inward looking autarkic industrial development became widespread in South Asia from 1950s till mid 1970s. Though South Asian countries (mainly four i.e. India, Pakistan, Bangladesh and Sri Lanka) were in transition in late 1970s, from inward looking policies to policies of relative outward orientation, they were not very successful in adopting these policies. Export promotion as one of their strategies did not help them in adjusting to the crises through increased export earnings. As a matter of fact, their market shares in world exports declined. They were outstripped by other developing countries who emerged more successful in gaining export market shares in addition to what they already possessed. Naturally, one becomes anxious to know why, South Asian countries, having such a vast potential of exports, huge network of institutions and with their efforts in continuous framing and reframing of import-export policies, have been surpassed by developed nations in the 1950s, resource-poor Japan in the mid

1950s, NIC-4 in the 1960s, ASEAN-4 in the 1970s and China in the 1980s.

The analysis of development performance of NIC-4, ASEAN-4 and South Asia (James, Naya and Meier, 1987)[1], relative progress of East Asia, South Asia, and Latin America (Naya, Seiji and others, 1989)[2], balance of payment adjustments of South Asia (Neela Mukherjee, 1987)[3], and survey of India's trade regimes (Bhagwati and Srinivasan, 1975)[4], reveals that the success of trade liberalisation of NIC-4 and ASEAN-4 lies in their combination of outward oriented trade policies and cautious macroeconomic policies aimed at eliminating market distortions to improve efficiency of resource use and the international competitiveness of exports.

In the case of South Asia the process of trade liberalisation has been gradual, slow and detrimental to the growth due to lack of competitiveness and inconsistency of macroeconomic policies to outward orientation till mid 1980s. Since mid 1980s four major South Asian countries (Bangladesh, India, Pakistan and Sri Lanka) have initiated structural economic reforms but lack of international competitiveness of several exportable products still seems to be the root cause of low growth of exports of South Asian countries. The success of the macroeconomic adjustments, for better performance of exports will depend on to what extent these reforms eliminate price distortions to improve the international competitiveness of exports.

Experience of successful exporting countries of the world shows that these countries improved their competitiveness with the development process. Two reasons may be attributed to this: (a) The competitiveness of exports may depend on whether composition of exports is determined by comparative advantage or not, and (b) The competitiveness is a dynamic phenomenon evolving as an integral component of the development process. The reasoning appears to be that sources of comparative advantage lay in factor endowments. Changes in factor endowments over a period of time (due to change in technology, scale economies etc.) will, therefore, lead to change in comparative advantage. A priori one would expect underdeveloped economies having scarcity of capital and skill to possess a traditional production structure and thus specialise in production and export of raw materials and agricultural products. As economies move from lower to higher levels of development, they accumulate capital, develop skills and generate better technology leading to changes in the structure of comparative costs. This ever-changing structure of comparative costs allows a given country to move up the ladder of comparative advantage from specialisation in primary products (i.e., raw materials and agricultural products) to manufacturing products which are labour-intensive in the first stage and then to manufacturing products which are capital, skill, technology and knowledge-intensive in subsequent stages.

Every low income economy, before development, has inadequate endowment of capital and thus possesses a traditional production structure that is intensive in natural resources and unskilled labour. Consequently, for every economy, the comparative advantage in the initial stage of development may be expected to be in raw materials and agricultural products. Thus growth model based on exports of primary and agricultural products in the initial stages of development has considerable advantages in maximising agriculture's contribution to overall development. "Higher rates of GDP growth in the ASEAN-4 were quite conducive to the expansion of agricultural exports while this was not the case in South Asia". "By expanding agricultural exports the ASEAN-4 countries realised greater foreign exchange earnings and improved the linkages between agriculture and other sectors". (James Naya and Meier, 1987) [5]. However, most of the South Asian countries had an active policy of promoting the exports of new manufactures and neglecting the traditional exports. The result was that neither the South Asian countries could maintain their traditional market shares nor could they capture any sizeable proportion of new markets. The past experience of export promotion of South Asia combined with strong international competition in manufacturing and external environment calls for a reassessment of the feasibility of continuing to rely on manufacturing export led growth as the major development dynamic in South Asia.

The South Asian economies, being developing economies, characterised by abundance of labour, scarcity of capital, relatively low level of technology and heavy dependence in natural endowments, their comparative advantage would seem to lie in agricultural products and labour intensive manufactures.

The present study is based on the fundamental premise that the South Asian economies, being developing economies, characterised by abundance of labour, scarcity of capital, relatively low level of technology and heavy dependence in natural endowments, their comparative advantage would seem to lie in agricultural products and labour intensive manufactures. An increase in the share of agricultural exports in the export basket of these countries may, thus, lead to better performance of their exports. There are several good reasons why agricultural exports should make a higher contribution to the total exports than the present level.

South Asian economies are basically agricultural in nature. Despite the rapid growth of industrial production, their dependence on agriculture is staggering. At present, about 50 per cent of national income of Sri Lanka, 40 per cent of the national income of India and 25 per cent of national income of Pakistan is derived from agricultural sector. There are several reasons why agricultural exports should contribute a similar proportion to total exports of these economies.

(a) In an economy where demand chronically exceeds supply, each sector should generate export supplies commensurate with its share of national output and productive resources". (Sengupta, 1986) [6].

(b) Domestic Resource Cost of each rupee of foreign exchange earned is lower in agriculture than in manufactures. (Sengupta, 1986) [7].

(c) Factor incomes generated at home per rupee of output exported, directly and through backward linkages, is higher for agriculture compared with manufacture and mining. (Sengupta 1986) [8].

(d) Basically, agricultural commodity exports require the least amount of cash assistance. Foreign exchange earning is net in most cases, as raw materials, capital equipment and components have not to be imported for agricultural production. (M.Dattatreyulu, 1988) [9].

(e) World demand for some of the higher value agricultural products such as meat, dairy items, marine products, fruits and vegetables and other processed and semi-processed agricultural products is increasing since last few years due to (i) increase in income of developing countries (ii) increasing health consciousness in western world leading to reviewed preference for natural and fresh foods.

(f) Along with the expected increased demand for agricultural, horticultural and marine products, South Asian countries enjoy enormous opportunities in the developing region especially in Asia and Africa because of similarity in socio-economic conditions, initial stages of their economic development and the geographical proximity to the region.

(g) South Asian countries enjoy a fairly large potential for agricultural, horticultural and marine products due to large geographical area with varied climate, temperatures and variety of soils accompanied by the longest coast lines in the world.

Despite the above advantages these countries' share in global exports of these products pales into insignificance. World trade in Agricultural commodities is of the order of about $300 billion and yet the share of South Asian countries is not more than 1 per cent. To look into the explanation for this paradoxical situation an analysis of South Asian countries, comparative advantage of trade in agricultural products has been attempted in this study. The objective is (a) to examine comparative advantage of South Asian countries in export of agricultural products and labour intensive products; (b) identify agricultural commodities where comparative advantage exists; (c) analyse their export performance; (d) suggest strategy for agricultural exports. However, due to vastness of the subject and data limitation, though the macro analysis i.e. analysis of sectoral composition of exports and comparative advantage will deal with South Asian economies in general, the micro analysis i.e. identification of agricultural producers having comparative advantage and their analysis will be done with special reference to India only.

1.1 Performance of Exports: Past Studies

The main purpose of this study is to work out the comparative advantage of trade in agricultural exports of South Asia. Thus in length the literature on agricultural export, literature on exports and growth, manufacturing export, factors determining exports and comparative advantage of exports was reviewed.

An enormous literature is available on exports, strategy or exports, export and growth and manufacturing exports, till recently, not much literature (except a few reports by FAO., ITC, UNCTAD/GATT, CECD and Asian Surveys by ADB) [10] was available on agricultural exports. A few of the latest publications on Agricultural exports are (a) Horticultural Exports of Developing Countries by Nurul Islam [11]; (b) Export prospects of cereals, fruits and vegetable by Sengupta [12]; and (c) Agricultural exports strategy by ESRF (Economic and Scientific Research Foundation) [13].

The study by Nurul Islam gives a comprehensive analysis of world trade in fruits and vegetables with special emphasis on exports of developing countries. The study suggests that a detailed individual country studies are needed to analyse the factors that affect the production and exportation of horticultural products in specific circumstances.

According to the study by Sengupta, "the potential role of agricultural exports, specially of the non-traditional items, has still not been fully assessed". Sengupta (1986) [14]. The emphasis in this study has been on assessing India's internal compulsions, strengths and constraints and relating them to the world market situation.

The study of ESRF(Economic and Scientific Research Foundation) firstly stresses the need to increase agricultural exports of India and then proceeds with empirical the analysis of agricultural exports.

Most of the studies on agricultural exports deal with performance, problems and prospects of agricultural exports. There seems to be no study which deals with comparative advantage in agricultural exports. As developing economies seem to have comparative advantage in agricultural exports, this study intends to deal with comparative advantage in Agricultural exports of South Asian countries.

The literature available on exports, strategy of exports, exports and growth, manufacturing exports and factors determining exports is so enormous that it may not be feasible to give a review of even part of the whole literature. We may attempt to give a gist of findings of the studies dealing mainly with factors determining exports.

There have been a number of studies to identify and assess the influence of factors underlying performance of exports of a particular country or a group of countries. Some of the studies worth mentioning are various reports of World Bank, U.N., F.A.O., Studies of Bela Balassa (1971; 1982, 1989), Literature on Comparative Advantage by U.N. (1982, 1983, 1985, 1986), studies of Leamer (1982, 1984), Asian surveys by Asian Development Bank (1969, 1976, 1982, 1987), Comparative study of East Asia, Southeast Asian and South Asian countries by James Naya and Meier (1987), Balance of Payments Adjustments of Southeast Asian and South Asian countries by James Naya and Meier (1987), Balance of Payments Adjustments of South Asian by Neela Mukherji (1987), Lessons in Development, A comparative study of Asia and Latin America by Naya, Urutia, Mark and Fuentes (1989) and studies on India exports by Deepak Nayyar (1980) Manmohan Singh (1966), Bhagwati and Srinivasan (1975) and Martin Wolf (1982) [15].

The above mentioned studies highlighted the influence and importance of (a) foreign demand for a country's exports, (b) commercial policies abroad, (c) domestic supply of exports, and (d) domestic policies towards exportables as the basic factors underlying export performance. Some studies highlighted the importance of external factors while others justifies that domestic factors were crucial determinants of export growth. Analysis of India's exports by Deepak Nayyar, R. Banerjee and Martin Wolf highlight the importance and influence of domestic factors on India's exports. Bela Balassa, James Naya and Meier and Naya, Seiji, Urutia M. Mark and S. Fuentes and certain studies by U.N. highlighted the importance of trade policies of an economy as the main determinant of exports. Recently, since the publication of some studies by Bela Balassa and U.N. on comparative advantage and trade and the use of Constant Market Share model for empirical analysis of export growth, the main cause of poor performance of exports is considered to be lack of competitiveness. The results of Constant Market Share analysis of India by Deepak Nayyar, R. Banerjee, Manmohan Aggarwal and NIC-4, ASEAN-4 and South Asian countries by U.N. show lack of competitiveness to be the main cause of poor performance of South Asian countries' exports as compared to NIC-4 and ASEAN-4.

Thus we may state that trade policies provide the main explanation of overall export performance.

It has been proved by economists that countries applying outward oriented development strategies had a better performance in terms of exports, economic growth and employment, whereas countries opting for inward orientation showed poor performance.

NICs opted for export orientation in beginning of 1960s. Exports in the NICs grew at an average rate of 20 to 30 per cent in 1960s and 1970s [16]. The ASEAN-4 countries shifted to export orientation by mid of 1960s and were able to expand exports in the

1970s. However, South Asian countries continued with inward-orientation till the end of 1970s.

Export promotion, as one of their strategies did not help them in adjusting to the crisis situation through increased export earnings. As a matter of fact, their market share in world exports declined over the seventies. They were outstripped by other developing countries who emerged more successful in gaining export market shares in addition, to what they already possessed, the reason being, inadequate consideration given to the resource endowment structure and thus comparative advantage in planning for industrialisation and trade.

Outward oriented trade policy will help optimise use of available factors of production only if pursued in right earnestness (or perspective). An exclusive focus on exports, leading perhaps to chaotic policies to promote them regardless of cost could result in misallocation of resources [17]. The studies of James Naya and Meier [18] and Naya, Urutia, Mark and Fuentes [19] reveal that international competitiveness was greatly emphasised in NIC-4 and ASEAN-4 trade regimes whereas trade regimes in South Asia led to distortions in trade [20].

Industries receiving protection were pushed rapidly to become internationally competitive; firms were expected to produce goods of comparable quality and price to imports, and to begin exporting within a few years. The requirement to produce at world market prices prevented proliferation of excessively capital intensive activities. The overall industrialisation strategy of the NICs supported an output mix that conformed to domestic factor endowments. The resources available had a strong influence on the pattern of industrial developments.

In most South Asian countries export promotion policies were inefficiently designed and implemented. They could best be characterised by a system of indiscriminate export promotion with even a perverse bias whereby the subsidy was inversely related to the competitive strength. Policies of these countries aimed at industrial self sufficiency (especially in basic industries e.g.

chemicals, metal products, machinery) under close supervision of the government and largely disregarded the country's factor endowments and the bottlenecks to industrial development (e.g., infrastructure and energy shortage). A whole range of subsidies and other incentives were offered to exports of manufactures. Most of these countries had an active policy of promoting the exports of new manufactures and neglecting the traditional exports. The result was neither South Asian countries could maintain their traditional market shares nor could hey capture any sizeable proportion of new markets. These economies, at present, confront crisis in their international competitive position.

(a) Though the pace of industrialisation and trade is affected by initial conditions - factor and resource endowment, size location and social moves, international environment, industrial and trade growth rates and structural change, government policy is the main determinant of industrial and trade growth rates. Trade policy is the main instrument to tailor trade to the needs of growth and development.

(b) Countries applying outward oriented development strategies had a better performance in terms of exports, economic growth and employment. Countries with, inward orientation showed poor performance.

(c) An outward oriented trade policy will help optimise use of available factors of production only if it is pursued in right perspective. An exclusive focus on exports, leading perhaps to chaotic policies to promote them regardless of cost could result in misallocation of resources. This may be the reason that whereas some of the Asian economies (mainly East Asia) could establish their name and image as exporting nations in the world market whereas others (South Asian) failed to do so. In spite of various export promotion measures exports of South Asian countries (specially India) have failed to register growth rate compatible to (a) other developing nations, and (b) to the amount of efforts and initiative taken by the country.

(d) The main causal factor underlying poor performance of exports of South Asian countries (specially India and Pakistan) seems to be lack of emphasis on international competitiveness.

Most of the studies on performance of exports, however, give stress on the factors which determine exports on the basis of past performance. No study explores in greater detail the causal relationship. For instance, poor performance of exports of present day developing countries is attributed , by most economists, to lack of competitiveness. No study goes into the factors that actually determine competitiveness in a present day developing economy. Deepak Nayyar does mention the factors which determine competitiveness and studies by James Naya and Meier and Naya, Seiji, Urutia, Mark and Fuentes gives some reference of the fact as to how NIC-4 and ASEAN-4 countries could improve their competitiveness of exports in 1970s and 1980s by specialising in primary and agricultural products in the initial stages of development, shifting to labour intensive exports in the second stage, and capital intensive in the third stage. No study, however, deals with factors that may lead to improvement of export competitiveness in South Asia.

We may state that competitiveness of an underdeveloped economy is going to be less in all products as compared to developed economies. Enhancing competitiveness may be termed as one of the major objectives of trade strategy for a developing economy. Thus, we have to see what are the major corrections through which competitiveness can be improved. An attempt is made in this study to suggest how South Asian countries can evolve competitiveness in world trade by adopting the theory of shifting comparative advantage in exports i.e. by adopting a growth model based on exports of primary products as the comparative advantage for developing and low income agricultural economies of South Asia seems to lie in agricultural products and labour-intensive manufactures.

1.2 Methodology

The methodology adopted for this study consists of two parts:
(a) For classification of agricultural and allied products.
(b) For analysis of competitiveness of exports.

1.2.1 Classification of Agricultural and Allied Products

1.2.1.1 Alternative Classifications: Two alternative methodologies may mentioned in classifying agricultural and allied products. The first is classification contained in the Handbook of International Trade and Development Statistics UNCTAD, U.N. and the second is Leamer's trade aggregates.

As per UNCTAD classification, agricultural commodities include:

(1) All food items comprising RITC Sec 0, 1, Div. 22 and Sec. 4.

(2) Agricultural raw materials comprising RITC (Sec.2-Div (22+27+28). In other words agricultural and allied products comprise of Sec.0,1,2 (less 27 and 28) and Sec.4.

Leamer's trade aggregates are formed from the sixty one two digit SITC commodity classes by a set of cross section regressions of the net export data on a list of eleven resources. The commodities having similar regression coefficients are aggregated. Leamer's ten aggregates comprise: two primary product aggregates (petroleum and raw materials) four crop aggregates (forest products, tropical agricultural products, animal products and cereals) and four manufactured aggregates labour intensive, capital intensive, machinery and chemicals). These ten aggregates are abbreviated as PETRO, MAT, FOR, TROP, ANL, CER, LAB, CAP, MACH and CHEM respectively. As per Leamer, four crop aggregates i.e. FOR, TROP, ANL and CER comprise agricultural and allied products. In terms of SITC classification these aggregates (i.e. agricultural and allied products) comprise:-

(a) Food and live animals chiefly for food (Sec.0)

(b) Beverages and Tobacco (Sec.1).

(c) Crude materials inedible oils (except fuels) excluding crude fertiliser and crude minerals and metalliferrous ores and metal scrap [Sec.2-Div (27+28)].

(d) Animal oils and fats (Sec.4)

(e) Wood and Cork Manufactures (Div.63)

(f) Paper, Paper board (Div.64)

(g) Animals n.e.s. (Div.94).

Thus agricultural and allied products as per Leamer's Trade aggregates comprise Sec.0,1,2 (less 27 and 28), Sec.4, Div.63,64 and 94.

The composition as per both the methodologies is almost the same except for an addition of Div. 63, 64 and 94 in Leamer's aggregate over and above the aggregates classified in U.N. handbook.

1.2.1.2 Classification Used in the Present Study: We shall be using, in our analysis, Leamer's composition to define agricultural and allied products as it is based on econometric analysis of cross section multiple correlations as well as cross section simple correlations between trade data and resource data. Moreover, Leamer's composition is same as FAO's composition of total agricultural exports except addition of Div.94 in Leamer's composition over FAO's composition. Hence, in terms of two digit classification, agricultural and allied products shall comprise of 00, 01, 02, 03, 04, 05, 06, 07, 08, 09, 11, 12, 21, 22, 23, 24, 25, 26, 29, 41, 42, 43, 63, 64, and 94.

Since the number of agricultural commodities that are traded internationally is enormous, no study with any time constraint could analyse each commodity. The study, therefore, confines itself to two digit SITC aggregates for macro analysis of agricultural exports (where comparative study of exports of South Asia vis-à-vis NIC and ASEAN is attempted) and to three digit SITC aggregation for micro analysis of agricultural exports of India.

1.2.2 Competitiveness of Exports

Competitiveness may be defined as "The degree to which a nation, under free and marked conditions, produced goods and services that meet the test of the international markets while simultaneously maintaining and expanding the real incomes of its citizens".

1.2.2.1 Methodology Used in General: Two approaches, viz. Constant Market Share (CMS) and Domestic Resource Cost (DRC) are generally used for analysing the competitiveness of exports.

An immensely simplified model that has been frequently used in studies on the export performance of individual countries is the CMS analysis of export growth. Its basic assumption is that other things being equal, a country's share in the world market for exports should remain constant overtime. The model, by a procedure of decomposing eliminates from a given country's actual export performance these changes that are accounted by size and the pattern of the world trade and attributes the residual portion to the changes in export competitiveness. In this method, an increase in a country's exports can be divided into four parts which are attributable to the general growth in world trade, changes in the commodity composition of its exports, changes in the market distribution of its exports, and changes in competitiveness.

CMS model can only be used to analyse the past. The model has no stochastic basis and hence cannot be used for econometric projections or for forecasting future changes in the market share. Secondly, the model does not establish any casual relationship. "It is far too aggregative and does not go into the factors that actually determine each of the four components" (Deepak Nayyar). It only decomposes the past export growth of a country into its different components and stops short of providing any explanations as to why the exports increased or decreased as they did. Thirdly, the conclusions derived from an analysis based on CMS model is valid only for the particular time period chosen, the level of commodity aggregation adopted and for the particular breakdown of markets. A different set of these parameters may produce different results and presumably different conclusion.

An alternative approach to measure international competitiveness of exports is Domestic Resource Cost (DRC). DRC criterion is a single period, social cost benefit indicator giving the domestic factor cost and shadow prices of generating a unit of value added at international prices. DRC measures the cost, in terms of domestic factors of production evaluated at shadow prices, of earning or saving a net unit of foreign exchange as a consequence of producing a particular output. An output should be produced if its DRC is less than the shadow price of foreign exchange. Despite acknowledged methodological reservation, DRC measurements have proved to be highly useful policy tools when market of shadow prices diverge. Activities with the lowest DRC ratios are normally assumed to exhibit the greatest comparative advantage.

According to Chenery (1965) a country has a comparative advantage in producing and exporting a commodity if the social opportunity cost of producing a unit of commodity -- the value of all factors of production in their best alternative employment, is less than the commodity's export prices.

$$F.E. + D < P.E.$$

Where

F = Cost of direct and indirect foreign inputs per unit of output (in foreign currency).

E = The cost of a unit of foreign exchange.

D = The opportunity cost of direct and indirect domestic inputs per unit of output (in local currency)

P = The international price of the product (in foreign currency).

A second definition of DRC, has been suggested and applied by Bruno to measure the comparative advantage of economic activities producing tradable goods and services within a single country. Bruno's DRC methodology for measuring the comparative advantage of different tradable products is a simple modification of Chenery's methodology

$$DRC = \frac{D}{P-F} < E$$

According to this definition a country has a comparative advantage in the production of an exportable commodity if the social opportunity cost of earning a unit of foreign exchange is less than the exchange rate.

An alternative application of the DRC technique has been used by Pearson and Meyer (1974). The Pearson and Meyer adoption is useful for cross country comparisons of relative comparative advantage in production of an identical commodity.

$$\frac{D/(P-F)}{E} < 1$$

A country's relative efficiency in the production of a commodity can be evaluated by comparing its (D/P-F)/E ratio with that of its potential competitors. The smaller the ratio, the greater a country's relative efficiency.

A major drawback of DRC approach (like CMS approach) is that it cannot provide an understanding of the historical forces which gave rise to the current pattern of measured comparative advantage, nor does it provide much insight into likely directions of change in the future. DRC ratios have little to say about dynamic, as opposed to static, comparative advantage, and there have been few attempts to examine changes in the DRC overtime within the same economy.

1.2.2.2 Justification for the Methodology Adopted in the Present Study: Both the approaches, however, have little to say about dynamic, as opposed to static, comparative advantage. Secondly, both approaches fail to establish any casual relationship as they are measures of competitiveness not the determinants. Competitiveness of an underdeveloped economy is going to be less in all products as compared to developed economies. It is to be evolved with development process. Thus we have to see what are the major corrections through which it could be improved. For this we have to provide an understanding of the historical forces which gave rise to the current pattern of measured comparative advantage and provide an insight into the likely direction of change in the future, i.e. we have to determine the sources of comparative advantage and further whether the composition of trade is determined by this comparative advantage (static as well as dynamic) or not.

1.2.2.3 Methodology Adopted in the Present Study: The Heckscher-Ohlin (HO) model has conventionally been used to explain the sources of comparative advantage. According to this theory (assuming the technology to be same across countries) comparative advantage is the result of the resource endowment structure of different economies. A series of modifications to this theorem have been developed over the last forty years. A detailed review of these modifications reveal that a strong HO component should be present among the determinants of composition of trade in developing economies of South Asia. Thus the methodology adopted is:

1. To determine ideal comparative advantage based on resource endowment structure, i.e. to see impact on comparative advantage by interaction between factor endowment and factor intensity.

2. A rigorous analysis of patterns of ideal comparative advantage, requires cost comparisons across countries. But the data required for such an analysis are rarely available. Thus an index of comparative advantage, i.e. Revealed Comparative Advantage (RCA) for different sectors is worked out for estimating comparative advantage of trade in South Asian economies, NIC-4 and ASEAN-4.

3. Sectoral composition of exports as per Leamer's trade aggregates is worked out for South Asia, NIC-4 and ASEAN-4.

4. To examine the extent to which South Asian countries pattern of trade conforms to the structure of comparative advantage (in static as well as in dynamic sense) an empirical

verification of South Asian countries pattern of trade is attempted. The static nature of export compatibility of South Asia is analysed by an application of the commodity version proposition of the HO theory. The dynamic nature of compatibility of India's exports is analysed by testing the Bela Balassa's theory of "The Stages Approach of Comparative Advantage". As the concept of comparative advantage involves inter-country comparisons a comparison between South Asia, NIC-4 and ASEAN-4 countries' composition of export vis-a-vis comparative advantage has been done.

5. To identify products having comparative advantage in agriculture, RCA index at three digit level was worked out for all the agricultural and allied products in India.

Notes

1. James, W.E., Naya, Seiji and Meier, G.M. (1987): Asian Development.
2. Naya, Seiji, Urutia, M. Mark and S. Fuentes (1989): Lessons in Development.
3. Mukherji, Neela (1987): BOP Adjustments and South Asian Economies.
4. Srinivasan, T.N. and Bhagwati, S.N. (1975): Foreign Trade Regimes and Economic Development: India.
5. James, W.E., Naya, Seiji and Meier, G.M. (1987): Asian Development.
6. Sengupta, S.K. (1986): Export Prospects of Cereals, Fruits and Vegetables.
7. Ibid.
8. Ibid.
9. Dattatreyulu (1988): India's Agricultural Exports.
10. Please See Bibliography for details.
11 Nurul Islam (1990): Horticultural Exports of Developing countries.
12. Sengupta, S.K. (1986): Export Prospects of Cereals, Fruits and Vegetables.

13. Economic and Research Foundation (1986): Agricultural Exports Strategy: Problems and Prospects.

14. Sengupta, S.K. (1986): Export Prospects of Cereals, Fruits and Vegetables.

15. Please see bibliography for details of references.

16. James, Naya and Meier (1986) : Asian Development.

17. Balassa, Bela (1982) : Development Strategies in Semi-industrial Economies.

18. James, W.E., Naya, Seiji and Meier, G.M. ibid.

19. Naya, Seiji, Urutia, M.S.S. : Lessons in Development and Fuentes, A.C. (1989).

20. Srinivasan, T.N. and Bhagwati, J.N. (1975) : Foreign Trade Regimes and Economic Development : India.

Chapter 2

Export Performance of South Asian Countries

2.1 Introduction

This chapter outlines descriptive approach for causal factors that could have led to poor performance of exports in South Asia as compared to South East Asia and East Asia. Explaining the causes of economic growth and backwardness in one's own region as compared to another region and attempting to compare development experiences often helps to clarify which variables and policies are strategic in the growth process.

The economic growth of the Asia Pacific region, which encompasses the Western Basin from Korea in the north to Indonesia in the south, has been most impressive in the past two decades (1970s and 1980s). As compared to overall growth in Gross Domestic Product of about 5.0 percent in the 1960s, overall Gross Domestic Product growth increased by 6.5 percent in the 1970s and 6.6 percent in the 1980s [1]. The region has emerged as a major exporter of edible oils and several agricultural raw materials. The principal impetus to growth was provided by rapid growth in manufacturing, which in turn was greatly facilitated by a relatively favourable environment for exports, especially during the 1970s.

Asian growth, however, has not been uniform. During 1970s and 1980s, some of the countries emerged as newly industrialising countries, some were in the process of entering this group while others lagged behind. It will not be wrong to classify the resulting groups into three main categories:

(a) The high income, newly industrialising countries (NICs) namely, HongKong, Singapore, Korea and Taiwan.

(b) The middle income emerging industrialising countries, namely, Indonesia, Malaysia, the Philippines and Thailand. These countries are commonly known as ASEAN-4 - the four larger members of the Association of South East Asian Nations.

(c) The low income countries of South Asia namely Bangladesh, India, Pakistan, Nepal and Sri Lanka.

2.1.1 Development Performance of the Three Subregions - An Overview

The Asian newly industrialising countries - Hong Kong, Singapore, Korea and Taiwan have achieved economic growth rates virtually without historical precedent, and the pace with which they have industrialised is amazing. Three resource-rich SouthEast Asian Nations - Indonesia, Malaysia and Thailand also demonstrate vigorous economic growth. Gross Domestic Product (GDP) rates for these countries were much higher than the average for less developed countries. However, average annual rates of growth of GDP have been lower in the South Asian countries than in the NICs and ASEAN throughout 1960s and 1970s. [Table 2.1]

2.1.1.1 South Asia: The South Asian countries have registered the slowest economic growth (Table 2.1). In the 1970s real gross domestic product increased by some 2 percent, 3 percent and 5 percent per year in Nepal, India and Sri Lanka respectively - rates marginally below those of 1960s. Bangladesh and Pakistan showed real gross domestic product growth of 3 percent and 5 percent respectively over the same period. South Asian agricultural production in the 1970s barely kept pace with population growth, while industrial production increased by less than 5 percent per year. Per capita income increase has been relatively low and poverty remains widespread. In the 1980s South Asia performed much better till mid-1980s, with annual growth increasing during the period 1980-86 in all the countries

Table 2.1: Growth of Real GDP and GDP per capita (compounded annual change)South Asia, ASEAN-4 and NIC-4

Year(→) Country(↓)	Real GDP					Real GDP per capita	
	1960-70	1970-80	1980-85	1986	1987	1960-85	1960-86
NICS							
Hongkong	10.1	9.8	4.2	8.7	13.6	6.3	5.9
Korea, Republic of	9.5	8.2	7.5	11.4	11.2	6.4	6.6
Singapore	9.2	9.1	6.0	1.9	8.8	6.7	6.4
Taiwan	9.6	9.7	6.1	9.9	11.2	6.4	6.5
ASEAN - 4							
Indonesia	3.8	8.0	3.6	2.5	3.5	3.0	3.1
Malaysia	6.5	7.9	5.1	0.5	4.7	4.2	3.7
Philippines (GNP)	5.2	6.3	-0.9	0.1	4.6	1.6	1.5
Thailand	7.9	6.9	5.3	3.8	6.6	4.2	4.0

SOUTH ASIA							
	N.A.	5.8[a]	3.9	5.2	4.0	2.5	2.5
Bangladesh	3.9	3.2	5.4	5.0	1.5	1.6	1.7
India	N.A.	5.6[a]	6.3	7.2	7.7	2.7	2.9
Pakistan	5.8	4.6	5.0	4.2	3.1	3.2	3.2
Sri Lanka							

N.A. = Not available.

a = 1973-80.

Source: (1) Asian Development Bank Annual Report, 1986, (2) Asian Development Bank (Key Indicators of Developing Member Countries of ADB, 1987 (3) International Monetary Fund, International Financial Statistics Year Book, 1987, 1988. (4) World Bank, World Development Report, 1983 and 1990.

except Bangladesh. (The major factor responsible for the improvement in growth rates was the excellent performance of the agricultural sector). However, in late 1980s there was major set back in South Asian economic growth due to severe drought in several countries in 1987 and floods in Bangladesh in 1988.

2.1.1.2 The Newly Industrialising Countries (NIC-4): Since the beginning of the 1960s, the newly industrialising countries have been the fastest growing countries in Asia and perhaps in the world. In the 1960s their gross domestic product growth averaged more than 9 percent a year and in the 1970s between 8 and 9 percent a year (Table 2.1). Their growth was low from 1980 to 85 because of the second oil price shock and depressed demand for their exports in industrialised countries. Since 1986, however, these countries have again shown rapid growth, with Singapore averaging 9 percent and other three newly industrialising countries more than 10 percent.

2.1.1.3 The Association of South East Asian Nations (ASEAN-4): During the 1970s, the performance of the Association of South East Asian Nations ASEAN-4, countries was quite impressive. The annual gross domestic product growth rate of these countries during the period ranged from about 6 percent for the Philippines to 8 percent for Indonesia (Table 2.1). Although industrial production increased at a fairly high rate, the major stimulus to growth came from the excellent performance of the agriculture and mineral sectors. This situation, however, changed dramatically in all of these countries during the 1980s. Between 1980 and 1986 the average gross domestic product growth rate of the ASEAN-4 was nearly half of that in 1970s. Since 1987, however, there was improvement in growth rate of these countries.

A brief overview of development performance of three subregions of Asia depicts that countries at a similar stage of development in 1950s showed varied economic performance during the next three decades. The economic performance of the Asian NIC-4, since 1960s, has been spectacular, surpassing that of any other group of countries in modern times. Almost the same can be said of the ASEAN-4 (except the Philippines). But the economic performance of South Asian countries and the Philippines has been less impressive.

2.2 Causal Factors for Differences in Growth Rates among the Three Subregions

Current interest and research in economic development has focused on examining the causal factors for differences in growth rates among developing economies. Some of the factors responsible for the poor performance of South Asian countries as compared to NIC-4 and ASEAN-4 are:

(a). Initial conditions of the economy.

(b). Small size of the country and lack of resources.

(c). Nature of resource development

(d). Fortuitous factors in the global scenario, including flow of foreign aid and a relatively unchallenged market for light manufactures.

(e). Heritage factor, i.e. the work ethics of some of the Asian countries which encourage discipline, hardwork and loyalty.

Lack of natural resources and small size of the country is considered to be the most important causal factor for differential growth rates in the three subregions of Asia. The economic explanation given for this causal factor is that abundance of natural resources leads to a comfortable balance of payment position and a strong national currency which discriminates against exports of manufactures. Countries with no natural resources, on the other hand, must make the most of their human resources. The economic explanation, however, is suggestive at best - some countries have done well with resources while others without resources. The newly industrialising countries used their large and skilled labour force and emphasised exports of labour intensive manufactures. The rapid growth of these industries led to increasing wages and improvement in income distribution. The declining terms of trade in the 1980s forced the ASEAN-4 countries to attempt to diversify their economic and export structures. Exports of light manufactures that use abundant labour have become increasingly important.

Several fortuitous events in the global scenario including international trading environment, large foreign aid, commodity prices and currency realignments are emphasised as important factors behind the newly industrialising countries' rapid economic growth. External factors, do indeed, play an important role. Yet, export growth as well as economic growth in newly industrialising countries continued in the 1970s despite turbulent events and slower world trade due to efficient trade policies of their governments. Moreover, it is difficult to measure separately the impact of different external factors in various countries as these factors, apart from influencing each other at the international level, affect many macro-economic variables of these countries.

The success of Asian countries can also be attributed to their Confucian heritage and other characteristics of Asian ethics that encourage hard work, loyalty and diligence. There can be no doubt that the quality of the labour force and entrepreneurial talent have played a major role. But these factors cannot be attributed solely to cultural heritage. The provision of opportunities for educational advancement and market activities (which in turn depend on Government policy) have also contributed to the development of human resources and institutions in the successful Asian countries [2].

Thus, if we draw on the wider evidence of the debate about the causes of development, voluminous literature on development suggests that though the pace of industrialisation and trade depends partly on initial conditions, factors and resource endowment, country size, location and social mores and on international environment, the principal cause of differentials in growth rate is policy formulation and implementation by the

Governments of different countries. In particular, trade policies of developing countries can affect trade as well as economic performance significantly. The rapid industrial growth of many of the Asian developing countries compared to other developing regions reflects the different types of policies they have adopted.

The two major approaches to industrialisation are frequently identified as import substitution i.e. inward orientation and export orientation i.e. outward looking development strategy.

2.2.1 Outward Oriented Trade Policies

Outward looking strategies have been shown to raise income whereas import substitution often leads to the entrenchment of business. Studies by Srinivasan (1974) [3], Bela Balassa (1978) [4], Michael Michaely (1977) [5], William B. Tyler (1981) [6] have found a significant positive relationship between the growth of manufactured exports and economic growth.

The superior performance of countries under export-oriented growth is spurred by a more efficient allocation of resources from participation in world trade. Traditional international trade theory explains this as being the result of specialisation in accordance with comparative advantage. Manufacturing for exports as well as for domestic consumption encourages efficient allocation of resources, permits the exploitation of economies of scale, generates technological improvements in response to competition abroad, and in labour surplus countries, contributes to increased employment. Furthermore, the strong growth and balance of payments positions that countries can attain by promoting exports enable them to borrow abroad to supplement domestic savings. By contrast, though an import substituting strategy is often appealing to developing countries thinking that replacing previously imported goods with domestic production will ease the balance of payment situation and at the same time fulfil the countries' aspirations towards industrialisation, it often leads to distortions and serious balance of payment problems in an economy.

Analysis of development performance of NIC's, ASEAN-4 and South Asian countries by James, Naya and Meier (1987) [7], relative progress of East, South-East South Asia and Latin America by Naya Seiji, M. Urutia, S. Mark and A. Fuentes (1989) [8] and Development Strategy of 11 semi-industrialising countries by Bela Balassa (1982) [9] reveals that countries applying outward oriented development strategies had a better performance in terms of exports, economic growth and employment whereas countries with continued inward orientation lead to increasing economic difficulties.

Most of the Asian developing countries followed protectionist policies to achieve rapid industrial and economic growth. Over the years this has significantly changed. The resource-poor NICs were the first to recognise the importance of liberal trade policies in ensuring efficiency and productivity. They shifted to export orientation by early 1960s. Korea and Taiwan moved in to export orientation when the first stage of import substitution (IS) was over (1951-57). HongKong did not even pass through first stage of import substitution (IS), and in Singapore the first stage was very short. The ASEAN-4 began to liberalise trade in the 1970s. However, a pattern of inward looking, autarkic industrial development became widespread in South Asia in the 1950s, 1960s and even early 1970s for some countries. Countries that maintain more liberal trade regimes tend to enjoy greater levels of trade measured in terms of export and import ratios to gross domestic product. Among the developing Asian countries, the high income NICs typically enforce the most liberal import regimes and enjoy export to GDP ratios in excess of about 50 percent. The middle income South East Asian countries enforce somewhat more restrictive import policies and accordingly enjoy export to GDP ratios between 30 percent and 40 percent. Finally, the low income central and South Asian countries enforce the most restrictive import policies and consequently enjoy export to

GDP ratios no greater than about 15 percent, except in the case of Sri Lanka which enforces relatively low tariff levels [Table 2.2].

In a nutshell, the key factor behind the success of NICs and ASEAN-4 is the outward looking development strategies adopted by these countries. These countries, following export-oriented policies, enjoyed higher rates of real economic growth and were able to release their foreign exchange constraint. South Asian countries which continued with import substitution suffered balance of payments difficulties.

Although South Asian economies were in transition in the 1970s, from inward looking policies to policies of relative outward orientation, they were not very successful in adopting such measures. The level of trade measured in terms of exports to gross domestic product did not show much improvement for these economies. Though export to GDP ratios improved for South Asian countries in the 80s, they were still quite low as compared to NICs and ASEAN-4 (Table 2.2).

2.3 Performance of Exports: South Asia

The rise in fuel prices (due to oil price shocks) and the general inflationary world situation raised the import bills of the South Asian economies during 1970s. This lent an adverse impact on their BOP situation which necessitated an intensification of export promotion efforts by these economies. Thus, South Asian countries (mainly four, i.e. India, Pakistan, Bangladesh and Sri Lanka) were in transition in late 1970's from inward looking policies to policies of relative outward orientation. A wide range of subsidies and other incentives were offered to exports of manufactures. Measures were taken to liberalise imports and increase industrial production for the promotion of exports. However, a whole range of subsidies and other incentives were offered to exports of manufacturers and there was negative subsidy in case of most of the agricultural exports. In other words we may say that most of these countries had an active

Table 2.2: Ratio of Exports to GDP, 1970, 1978, 1985 and 1986 (percent at current prices) South Asia, ASEAN-4 and NIC-4

Year (→) Country (↓)	Exports			
	1970	1978	1985	1986
NIC-4				
Hong Kong	92.9	84.6	105.8	113.3
Korea(ROK)	14.0	30.3	36.4	40.9
Singapore[a]	81.9	128.9	124.8	129.8
Taiwan	29.7	52.4	55.1	60.4
ASEAN-4				
Indonesia	12.8	21.7	22.7	20.8
Malaysia	46.1	49.1	54.9	56.8
Philippines	19.1	18.2	20.8	24.7
Thailand	16.7	21.5	26.3	28.2

SOUTH ASIA				
Bangladesh	6.0[b]	5.5	6.5	6.5
Burma	5.6	5.8	-	5.0
India	4.4	7.3	6.8	6.6
Nepal	4.9	5.6	10.9	6.0
Pakistan	7.6	9.4	26.0	11.7
Sri Lanka	25.5	34.8	-	23.7
China[d,e]	2.9	5.6	-	13.9

Source : (1) Asian Development Bank (Key Indicators of Developing Member Countries) of ADB, July 1987, (2) International Monetary Fund, International Financial Statistics year book 1987 and 1988.

policy of promoting the exports of new manufacturer and neglecting the exports of agricultural products and traditional manufacture exports. The result was, being weak competitors in the world market for manufacturers, neither could they maintain their traditional market shares nor could they capture any sizeable proportion of new markets emerging in Gulf countries and the South East Asia. These economies, at present, confront a difficult situation in their international competitive position. In spite of the efforts to diversify their exports (along regional and commodity dimensions) and the introduction of numerous export promotion measures, their exports failed to register growth rates either comparable to those of other developing nations or commensurate with the amount of effort and initiative taken by the country. Export promotion as one of their strategies did not help them in adjusting to the crisis through increased export earnings. We hardly need to reiterate that their market shares in world exports declined (Table 2.3). They were outstripped by other developing countries (NIC-4 and ASEAN-4) who emerged more successful in gaining export market shares in addition to what they already possessed.

Thus while world trade has been growing at a phenomenal pace, we find that South Asian countries could not keep pace with this increase. Undoubtedly a large proportion of this expansion in world trade is attributable to increased exports from the developed countries and it might be argued that developing countries of South Asia could not have possibly competed with the developed countries and maintained its share in world exports. But how does one account for the fact·that South Asian countries exports did not even keep pace with the general growth of exports from the less developed countries? Naturally one becomes anxious to know why, South Asian countries, having such a vast potential for exports, huge network of institutions and with their efforts in continuous framing and reframing of import-export policies, have been surpassed by developed nations in the

Table 2.3: Percentage Share of Exports of Selected Countries in Total World Exports

Year	Korea	Taiwan	Hongkong	Singapore	Indonesia	Malaysia	Philipines	Thailand	India	Pakistan	Sri Lanka
1960	0.278	0.139	0.580	1.703	0.703	0.999	0.525	0.346	1.121	0.332	0.324
1965	0.101	0.260	0.661	0.956	0.410	0.715	0.404	0.360	0.976	0.306	0.237
1966	0.132	0.283	0.699	0.568	0.350	0.663	0.436	0.358	1.031	0.318	0.189
1967	0.161	0.323	0.769	0.574	0.335	0.613	0.413	0.343	0.813	0.325	0.175
1970	0.288	0.493	0.068	0.536	0.302	0.582	0.359	0.245	0.699	0.137	0.110
1971	0.327	0.612	0.831	0.540	0.379	0.504	0.337	0.255	0.623	0.143	0.105
1972	0.418	0.742	0.884	0.563	0.457	0.443	0.283	0.270	0.629	0.178	0.987
1973	0.597	0.911	0.939	0.676	0.595	0.565	0.349	0.290	0.540	0.177	0.076
1974	0.554	0.685	0.741	0.722	0.922	0.526	0.339	0.304	0.400	0.138	0.065
1975	0.612	0.639	0.726	0.643	0.856	0.463	0.277	0.266	0.525	0.127	0.073
1976	0.819	0.865	0.988	0.699	0.907	0.562	0.271	0.316	0.589	0.124	0.060
1977	0.940	0.975	0.980	0.771	1.816	0.569	0.293	0.327	0.597	0.111	0.070
1978	1.029	1.026	0.927	0.828	0.942	0.600	0.275	0.331	0.540	0.119	0.068
1979	0.955	1.020	0.960	0.903	0.989	0.703	0.290	0.336	0.495	0.138	0.062
1980	0.922	1.043	1.041	1.021	1.155	0.683	0.303	0.343	0.452	0.138	0.056
1981	1.140	1.206	1.170	1.124	1.194	0.631	0.303	0.377	0.445	0.155	0.059
1982	1.260	1.273	1.212	1.199	1.286	0.694	0.207	0.401	0.540	0.138	0.094

1983	1.453	1.492	1.306	1.290	1.250	0.840	0.291	0.379	0.544	0.183	0.063
1984	1.640	1.787	1.509	1.350	1.229	0.930	0.296	0.416	0.558	0.143	0.031
1985	1.645	1.690	1.670	1.262	1.028	0.954	0.255	0.394	0.510	0.152	0.072
1986	1.744	1.997	1.780	1.130	0.745	0.691	0.240	0.445	0.477	0.170	0.061
1987	2.012	2.298	2.070	1.225	0.732	0.766	0.241	0.498	0.486	0.178	0.058
1988	2.253	2.246	2.344	1.459	0.723	0.784	0.261	0.583	0.494	0.168	0.055

Source: International Financial Statistics, IMF, (Various Issues).

1950s, resource poor Japan in the mid 1950s, NICs in the 1960s, ASEAN-4 in the 1970s and China in the 1980s.

2.4 Reasons for Poor Performance of Exports: South Asia

Broadly speaking four hypothesis have been suggested by economists to explain the factors determining exports in general.

(a). Stagnant world demand,

(b). Traditional commodity composition,

(c). Traditional Market Distribution, and

(d). Slow growth in the production of exportables in relation to domestic absorption of these products.

According to Martin Wolf (1982) [10], the remarkable success of some of the Asian developing countries achieved during the 1960s and 1970s weakens the force of the first hypothesis as a tenable reason for poor export performance of South Asian countries exports.

Traditional export composition cannot be a complete explanation for poor performance as other Asian countries did much better than South Asia with similar initial export composition in the initial stages of development. Every economy, before development, has the traditional composition and direction of trade, almost same as that of South Asia. South Asian countries' position was not different from that of the United States in the 1850s, with its dependence on cotton, or that of Japan in the 1950s, with its dependence on cotton textiles, that of NIC-4 in early 1960s with their dependence on agriculture and primary commodities and ASEAN-4 in 1970s with their dependence on agricultural and primary commodities. However, we find that even after the increase in proportion of manufactures in India's and Pakistan's exports (Table 2.4) relative to the earlier periods, export performance has not shown any improvement. Traditional composition and direction in the initial stages of

Table 2.4: Structure of Merchandise Exports/Imports of South Asian Countries 1965 and 1988

		Exports			Imports		
		Foods, Minerals Metals	Other Primary Commodities	Total Manufactures	Foods, Minerals Metals	Other Primary commodities	Total manufactures
India	1965	10	41	49	27	14	59
	1988	9	18	73	18	9	73
Pakistan	1965	2	62	36	23	5	72
	1988	1	30	69	33	13	53
Bangladesh	1965	NA	NA	NA	NA	NA	NA
	1988	2	29	69	33	9	58
Sri Lanka	1965	NA	NA	NA	49	4	46
	1988	7	50	43	34	3	64

Source : Regional Economic Cooperation Between SAARC and ASEAN and Selected Other Countries of Asia Pacific Region - Charan D.Wadhwa.

development should be regarded not as a constraint but as a challenge as diversification out of older categories into new ones is exactly the task of development which is to be met by the efficiently designed and implemented trade policies (Table 2.4).

As far as the fourth hypothesis is concerned, according to Martin Wolf, it is most complex and important argument to deal with. It assumes that exports are the result of unwanted surpluses and not of efficient specialisation. It may be noted that development process and trading opportunities are inter-dependent. Trade policies lead to specialisation in trade, specialisation leads to increase in production and increase in production (if accompanied by increased productivity) in turn will help improve competitiveness and hence increased exports. South Asian countries wished to expand exports while thwarting the ability to specialise in trade. Since development policy and output structure have been determined largely independently of trading opportunities in these countries, output necessarily appears as a constraint as far as exports are concerned. Thus the hypothesis of the production constraint does not explain the slow growth of exports. On the contrary it is a self fulfilling prophecy.

In a nutshell, traditional composition (both commodity and direction) and lack of production should be regarded not as a constraint but as a challenge. Diversification out of older categories into new ones, stagnant markets into expanding markets and increase in production is exactly the task of development. This task is to be met by efficiently designed and implemented trade policies. Trade policy is an instrument to improve competitiveness of exports in world market. Trade policy is partly a matter of selecting and implementing a strategy and overall approach, to take full advantage of potential benefits, limit the disruptions and cultivate the learning opportunities taking into account a country's particular situation and resources.

The analysis of export performance by James, Naya and Meier (1987) Bhagwati and Srinivasan (1975) [11] ESCAP (1985) [12] Neela Mukherjee (1988) [13], Naya, Urutia, M. Mark and S.

Fuentes, A. (1989) [14] and Deepak Nayyar (1976) [15] reveals that the success of trade liberalisation of NICs and ASEAN-4 lies in their combination of outward oriented trade policies and cautious macro-economic policies aimed at eliminating market distortions to improve efficiency of resource use and the international competitiveness of exports. In case of South Asia an exclusive focus on exports, leading perhaps to chaotic policies to promote them regardless of cost resulted in misallocation of resources.

The reason may be that though exports provide advantage over import substitution by contributing to resource allocation as per comparative advantage, greater capacity utilisation, the exploitation of economies of scale and improvements in technology, export expansion will help optimise resource allocation only if pursued in right perspective. An exclusive focus on exports leading perhaps to chaotic policies to promote them regardless of cost could result in misallocation of resources. In other words we may say that while export orientation is a better strategy than import substitution, the realisation of rapid growth under export-oriented industrialisation, however, cannot be achieved through automatic adjustments. The role of the Government has been important in correct pricing, in setting objectives and in implementing policies. For example, monopolies have to be destroyed, labour and market imperfections have to be removed, overvaluation of exchange rates has to be corrected, trade restrictions have to be reduced, and incentives for exporting have to be established. According to James, Naya and Meier [16]: "Trade liberalisation cannot be effective in improving exports and economic growth when macro economic policies are inconsistent and destabilising. Yet, even if macroeconomic stabilisation is achieved, trade reform may not be immediately effective. Export markets have to be developed and contested with competitors. There are risks involved in exposing domestic industry to international competition. In moving toward a more export-oriented outward looking strategy, each government must consider, within its own institutional context how best to proceed".

The diversity of national resources and international marketing situations means that a single policy prescription would not be useful. Each country is unique and policies should be adopted to the particular circumstances of each. Nevertheless it would be equally wrong to imply that no general lessons can be drawn from past experiences. There are some common elements in the strategies of developing countries that have been most successful in promoting exports.

2.5 Factors Supporting Outward-oriented Trade Policies

The outward looking development strategy adopted by the NICs involved more than export orientation, though openness to world trade and investment was a crucial element. The policies integrated national economies into the world market. Along with export promotion, they have generally adopted prudent macro-economic policy to control budget deficits and restrain excessive demand and inflation. They have generally been very successful at mobilisation of domestic resources and for most part relaxed the savings constraints to development. Prudent demand management has gone hand in hand with promotion of financial development.

The policy changes in Korea took place in 1960 and 1961 and in Taiwan from 1958 to 1960. Like any package of policies aiming at export promotion, the measures taken by the Korean and Taiwanese government included reforms in exchange rate systems, import liberalisation and export incentives. Firstly the multiple exchange rate system was gradually collapsed into a single rate system, and the exchange rate applicable to the bulk of imports and exports by private enterprises was devalued. Second, the government gradually liberalised and finally abolished the commodity import quota system. Import controls were also liberalised. Thirdly, balanced incentive system was provided for expansion of exports. Korea and Taiwan continued to provide protection as they moved toward an export-oriented growth path. Further, the incentive system was not biased against primary activities and within the primary sector there was little discrimination against exports.

The ASEAN-4 countries following NICs, according to James Naya and Meier (1987), [17] undertook similar reviews of their trade and industrial policies in 1970s. In the second half of 1970s, in an effort to adjust to the increases in the oil import bill, the governments of the Philippines and Thailand tried to increase manufactured exports. Export controls were relaxed, export taxes abolished or reduced. Duty free importation allowances were granted to exporters for necessary inputs. Export credit facilities were provided to exporters. Overall simple average tariff rates decreased sharply in the Philippines. The oil exporters, Indonesia and Malaysia, undertook similar reviews of trade and industrial policies in the late 1970s". Malaysia experienced substantial increase in the exports of some primary commodities that were subject to practically no discrimination. The Philippines also made gains in sugar exports under preferential arrangements with the US. It increased its share in the declining world exports of abaca and expanded its timber exports.

The success of the macro-economic adjustments, for better performance of exports will depend on to what extent these reforms eliminate price distortions to improve efficiency of resource use and the international competitiveness of exports.

An important feature of protection in the NICs was that, for the most part, international competitiveness was greatly emphasised. This meant that domestic prices were not allowed to exceed international prices for prolonged periods. In fact industries receiving protection were pushed rapidly to become internationally competitive, firms were expected to produce goods of comparable quality and price to imports, and to begin exporting within a few years. When export targets were not met measures were taken to rectify the situation, ranging from threats of sanctions to provision of additional incentives and government action to remove bottlenecks. The requirement to produce at world market prices, prevented proliferation of excessively capital intensive activities.

The overall trade and industrialisation strategy in NIC-4 supported an output mix that conformed to domestic factor endowments and hence comparative advantage. As factor endowments shifted with capital accumulation, NICs shifted their industrial structures and exports from agricultural and primary products toward the more skill intensive and capital intensive production and exports. Policies in the ASEAN-4 were quite conducive to expansion of agricultural exports. By expanding agricultural exports, the ASEAN-4 countries realised greater foreign exchange earnings.

In a nutshell, the combination of outward oriented trade policies and cautious macro economic management aimed at improving efficiency of resource use and the international competitiveness of exports gave the NICs and ASEAN-4 a great deal of flexibility.

In most South Asian countries, policy produced price distortions and direct government interventions adversely affected industrial development. These policies aimed at industrial self-sufficiency (specially in basic industries e.g. chemicals, metal products, machinery) under close government supervision and largely disregarded the countries' factor endowments and the bottlenecks to industrial development'. Inefficient public enterprises, distorted prices and bureaucratic rigidities have led to high cost and low productivity in the industry. The overall industrialisation strategy followed in South Asia has placed heavy demand on scarce factors (skills and capital) and has made little use of unskilled labour and land. A whole range of subsidies and other incentives (in South Asia) were offered to exports of manufactures. Most of these countries had an active policy of promoting the exports of new manufactures and neglecting the agricultural exports and traditional exports. Major agricultural/primary exports and traditional manufactured product exports were subject to export taxes in most of the South Asian countries.

Further, till beginning of 1980s, the exchange rates of a majority of these countries appreciated relative to their major trading partners. Price controls has been most extensive in the South Asian countries where some price controls are still in force. (The exception being Pakistan). Despite their remarkable domestic resource mobilisation, investment in South Asia has continued to exceed saving. This has made these countries dependent on external assistance which had led to difficult debt service problems.

Thus, in the case of South Asia, though most of the countries shifted to relative export orientation policies since late 1970s, the process of trade liberalisation has been gradual, slow and detrimental to growth due to lack of competitiveness and inconsistency of macro economic policies to outward orientation till mid of 1980s.

Since mid of 1980s four major South Asian countries (Bangladesh, India, Pakistan and Sri Lanka) have initiated structural economic reforms with balance of payment support from the IMF. The macro economic measures undertaken by these economies include reforms in exchange rate systems, import liberalisation and export incentives. An important change has been the replacement of quantitative controls with tariffs.

However, given the staggering trade deficits of the last several years, heavy debt burden and slow growth in productivity, reliance on macro economic policy alone will prove risky. Reliance on exchange rate corrections, runs the risk of higher inflation rates and/or recessionary conditions in South Asia as export promotion polices are still inefficiently designed and implemented. Greater stress has been placed on promoting exports through provision of incentives. The quality of these export products remains low and there have been no incentives to improve it. High import duties in many countries still make the domestic market more attractive than the import market. Lack of international competitiveness of several exportable products still seems to be the root cause of low growth of exports of South Asian countries.

Thus the combination of outward oriented trade policies and cautious macroeconomic management aimed at improving efficiency of resource use and the international competitiveness of exports gave the NICs and ASEAN-4 a great deal of flexibility, whereas in South Asia, trade liberalisation has been detrimental to export growth due to lack of competitiveness.

The above view point is strengthened by the empirical result of decomposition of growth in exports of ESCAP countries by U.N. [18].

The results of analysis using the constant market share model are presented in Table 2.5.

The results of constant market share analysis (Table 2.5) show - for HongKong, India and Pakistan, the contribution of commodity composition was positive, implying that their exports comprise commodities whose demand was growing relatively faster in world markets. For almost all the countries except the Asian least developed countries, the contribution of market distribution was positive For China, HongKong, Indonesia, Malaysia, the Philippines, the Republic of Korea, Singapore, Taiwan province and Thailand, the contribution of market distribution to total change is less than 50 percent while for India it is 65 percent and for Pakistan 342 percent. The price competitiveness residual seems to be very high in a positive sense for Indonesia, Korea and Taiwan but negative for India and Pakistan. It is well-known fact that Taiwan and Korea had managed to export more because their exports were competitive and Indonesia and Thailand to a lesser extent have become price competitive. But India and Pakistan's exports became non-competitive in the world. If we rank these results according to

Table 2.5: Constant Market Share Analysis of Changes in Exports by Countries 1965-1980

Country	Changes in exports due to				Changes in exports between 1965-67 & 1978
	Increase due to world trade	Market distribution	Commodity composition	Price competitiveness	
Korea	7.1	1.63	0.32	90.94	9101.1
Taiwan	18.06	11.12	-1.46	72.28	10443.1
HongKong	49.62	25.09	17.26	8.04	5845.5
Singapore	50.83	39.23	-10.47	20.41	6866.5
Malaysia	743.06	27.89	-18.39	17.43	4022.3
Indonesia	44.14	22.94	-17.72	50.65	7014.7
Thailand	58.80	0.59	-15.07	55.68	2833.4
Philippines	203.15	46.94	-68.31	-81.78	1314.9
China	67.83	33.25	-5.02	3.94	5792.4
India	154.16	64.76	3.72	-122.64	2329.1

Source : U.N., ESCAP (1985) ; Development Paper No.5.

ordinal analysis (Table 2.6), price competitiveness can be given the highest rank for Indonesia, Korea and Taiwan. For HongKong, the increase due to world trade gets the highest rank, market distribution next, commodity composition third and the price competitiveness residual the last. For India and Pakistan, increase due to world trade gets the first rank, market distribution second, composition third and competitiveness not only the last, but negative too. Since data for Pakistan for 1965-67 includes Bangladesh, the whole decomposition analysis was repeated for all the countries including Pakistan, using 1972 as the base year and 1980 as the terminal year.

Ranking of decomposition results (Table 2.6) revealed that even if we use 1972 as the base, price competitiveness still ranked first for Korea and Taiwan. It improved for Malaysia, Philippines, Singapore and Thailand but for India and Pakistan it was still the fourth. This may be because the ASEAN-4 countries shifted to outward-orientation in late sixties whereas NIC-4 in the beginning of 1960s. It was only after adopting the outward orientation policies that countries could take advantage of international trade. Though South Asian countries shifted to relative export orientation policies since late 1970s, intensified their export promotion efforts and introduced structural reforms since mid of 1980s, they have not been able to improve their position in international trade due to lack of competitiveness in exports.

The analysis of India's trade regimes by Srinivasan (1974) [19] are in line with the above results. According to Srinivasan - "India's foreign trade regimes in conjunction with domestic licensing policies in the industrial sector, led to economic inefficiencies and impaired her economic performance. This conclusion follows not merely from the static analysis, but also from the analysis of growth effects. The policy framework was detrimental on balance to the growth of the economy by wasteful interindustrial and interfirm allocation of resources, by permitting and encouraging expansion of excess capacity and by blunting

Table 2.6: Ranks of each of the components contributing to export growth by countries and areas for the period 1965-67 to 1978-80 and 1972-80 decomposition analysis

Country	Changes in exports due to							
	World trade		Market distribution		Commodity composition		Change in price competitiveness	
	'72-80	'65-67 to '78-80	'72-80	'65-67 to '78-80	'72-80	'65-67 to '78-80	'72-80	'65-67 to '78-80
Korea	2	2	4	3	3	4	2	1
Taiwan	2	2	3	3	4	4	2	1
Hong Kong	2	-	-	-	-	-	-	-
Singapore	2	1	3	2	4	4	2	3
Malaysia	1	1	3	2	4	4	2	3
Thailand	2	1	4	3	3	4	1	2
Indonesia	1	2	2	3	3	4	1	2
Philippines	1	1	3	2	4	3	2	4
China	2	1	4	2	3	4	1	3
India	1	1	3	2	2	3	4	4

Source : ESCAP, Development Paper No.5, 1985.

competition and hence the incentives for cost consciousness and quality improvement. Export promotion policies were inefficiently designed and implemented. They could best be characterised by a system of indiscriminate export promotion with even a perverse bias whereby the subsidy was inversely related to the competitive strength" (Srinivasan, 1974).

2.6 Conclusions

In a nutshell we may state that though most of the countries in South Asia shifted to relative export orientation policies since 1970s and have initiated structural economic reforms since mid of 1980s, lack of international competitiveness still seems to be the roof cause of poor performance of exports in these economies. Thus, the process of trade liberalisation which has been gradual, slow and detrimental to growth due to lack of competitiveness may be made easier, quicker and more successful by adopting trade policies which are aimed at improving the underlying competitiveness of exports of South Asian countries. What is needed to make the position less costly is the development of a national competitiveness policy and the major objective of this trade strategy may be termed as enhancing competitiveness of exports.

Notes

1. Source : Asian Development Bank, 1987.
2. James, W.E., Naya, Seiji and Meier, G.M. (1987): Asian Development.
3. Srinivasan (NBER, 1974) : Trade Strategies for Economic Development, The Asian Experience.
4. Balassa, Bela (1978) : Export Incentives and Export Performance in Developing Countries.
5. Michael Michialey (1977) : Export and Growth - An Empirical Investigation.

6. Tyler W.B. (1981) : Growth and Export Expansion in Developing Countries.

7. James, W.E. Naya, Seiji and Meier, G.M. (1987) : Asian Development.

8. Naya Seiji, M. Urutia S. Mark and A. Fuentes (1989) : Lessons in Development.

9. Balassa, Bela (1982) - "Development Strategies and Economic Performance: A Comparative Analysis of Eleven Semi Industrial Economies.

10. Wolf, Martin (1982) : India's Exports.

11. Bhagwati, Jagdish and Srinivasan, T. (1975): Indian Trade Regimes and Economic Development.

12. United Nations (1985) : Development Paper No.5, ESCAP.

13. Mukherji, Neela (1987) : BOP Adjustment and South Asian Economies.

14. Naya, Seiji, Urutia, M. Mark and S. Fuentes, A. (1989): Lessons in Development.

15. Nayyar, Deepak (1976) : India's Exports and Export Policies in 1960s.

16. James, W.E., Naya, Seiji and Meier, G.M. (1987) - Asian Development.

17. James, W.E., Naya, Seiji and Meier, G.M. (1987): Asian Development.

18. United Nations (1985) : Development Paper No. 5, ESCAP.

19. Srinivasan T.N., (NBER, 1974) ; India's Trade Regimes.

Chapter 3

The Determinants of Competitiveness

3.1. Introduction

It was stated in the previous chapter that South Asian countries, despite outward orientation in trade during late 1970s and macro economic adjustments in late 1980s, could not improve their position in international trade as compared to other developing Asian countries like Korea, Taiwan, Malaysia, Philippines, Indonesia and Thailand. The main causal factor underlying poor performance of exports of South Asian countries seemed to be lack of emphasis on international competitiveness in these economics as compared to NIC-4 and ASEAN-4. South Asian economies lagged behind in the international competition by undertaking the production and export of commodities, keeping in view the world demand only while disregarding country's particular situation and resources. On the other hand, NIC-4 and ASEAN-4 moved up the ladder of competitiveness in world trade by selecting and implementing a trade strategy to take full advantage of the potential benefits, taking into account country's particular situation and resources. Thus enhancing competitiveness may be termed as one of the major objectives of trade strategy for South Asia. A general question therefore arises: what are the major corrections through which competitiveness could be improved? In an attempt to answer this question, the analysis of the present chapter is divided into three parts:

(a) Determinants of competitiveness,

(b) Sources of comparative advantage i.e., theories of comparative advantage and empirical models review, and

(c) Theoretical validity of Heckscher-Ohlin model of trade.

3.2 Determinants of Competitiveness

Competitiveness is defined as "the degree to which a nation under free and market conditions, produce goods and services that met the test of the international markets while simultaneously maintaining and expanding the real incomes of its citizens"[1].

The competitiveness is a dynamic phenomenon evolving as an integral component of the development process. This is because the development process and trading opportunities are interdependent. Trade policies lead to specialisation in trade, specialisation in trade in turn leads to increase in production. Increase in production (if accompanied by increased productivity) will help improve competitiveness. Two reasons may be attributing to this logic:

Firstly, as reallocation of resources depends on comparative advantage, the types of goods that less developed countries are best able to produce and export should be based on comparative advantage considerations. The export oriented approach is, to a considerable extent based on the theory of international comparative advantage. Its supporters advocate specialisation in production and trade, using relative efficiency criteria. The experience of many developing countries of South Asia has clearly demonstrated that adequate consideration is not always given to comparative advantage of a country in planning for industrialisation and trade. The neglect of comparative advantage very often leads to diseconomies of scale (less than minimum efficient scale) and inefficiency in production. The strategy of industrialisation that takes comparative advantage into consideration provides a suitable basis for an international division of labour between countries of different levels of economic development. An international division of labour on this basis is considered to be a desirable objective in the long run interest of a harmonious development of the international trade and the world economy. Thus the competitiveness of exports may depend on whether the composition of exports, as suggested by trade theories, is determined by comparative advantage or not.

Secondly, in a rapidly changing world of capital accumulation and technological progress, comparative advantage itself has come to be regarded as a constantly changing (a dynamic) phenomenon. There might be changes in comparative advantage with the progress of economic growth. The reasoning appears to be that sources of comparative advantage lay in factor endowments. Every economy, before development, depends on natural endowments. Thus, for every economy, the comparative advantage, in the initial stages of development, lay in raw materials and agricultural products. With technological improvements the relative endowment of capital changes. With increased accumulation of capital, as compared to natural resources and labour, the comparative advantage shifts from agricultural and raw material products to labour intensive manufactured products. The development process is further initiated in a chain reaction. Accumulation of capital leads to better skill and technological advancements which further shifts the comparative advantage from labour intensive manufactures to capital intensive manufactures. This in turn leads to economies of scale, increase in the size of market etc. There is further increase in accumulation of capital leading to still better skills and technology and hence shift in the comparative advantage to skill and technology intensive manufactures. In accordance with this shift in comparative advantage, the composition of exports shifts from raw material and agricultural products to labour intensive manufactures, and ultimately to highly skilled and technology intensive manufactures.

In other words, the composition of a country's export changes with the growth of its economy. The pattern approach to development analysis by Clark, Kuznets, Watanabe, Shisido Chenery and Taylor and subsequently followed up by many others, has greatly enhanced our understanding of the way the production structure "normally" changes in response to economic growth. Chenery and Taylor have demonstrated that variations in the production structure are related systematically not only with variations in national income and country size but also with trade patterns (Banerjee, 1975) [2]. The pattern of trade as per trade theories may vary from country to country because of differences in their endowment with productive resources and skill, in scale economies and in technology. A priori one would expect countries which are at lower levels of economic development, possessing simple technology, having a scarcity of physical capital and skill but with plenty of labour supply to specialise in the production and export of manufactures that are simple to produce and do not require much skill and capital. Conversely, countries with advanced technology, well endowed with capital and skill and having a relative shortage of labour are expected to specialise in research and capital intensive manufactures. As countries move from lower to higher levels of development, they are likely to accumulate capital, develop skills and generate a better technology leading to changes in the structure of comparative (relative) costs. This ever-changing structure of comparative costs allows a given country to move up the ladder of comparative advantage from specialisation in primary product exports to unskilled labour intensive exports to capital intensive exports and further to knowledge and skill intensive exports.

3.2.1 Hypotheses

We may state that the competitiveness of an underdeveloped economy is going to be less in all products as compared to a developed economy. It is to be evolved with the development process. Enhancing competitiveness may be termed as one of the major objectives of trade strategy for a developing economy. Experience of successful exporting countries of the world show that these countries improved their competitiveness with the development process.

The diversity of national resources and international marketing situations means that a single policy description would not be useful. Each country is unique and policies should be adopted to the particular circumstances of each. Nevertheless it would be equally wrong to imply that no general lessons can be drawn from past experiences. There are some common elements in the strategies of developing countries that have been successful in enhancing competitiveness of exports. Thus, the process by which Japan, NIC-4 and ASEAN-4 evolved this competitiveness may serve as a good example for major corrections to be incorporated in South Asia.

Every economy before development has a traditional production structure, based on its natural resources with acute lack of capital. For every economy the comparative advantage in the initial stages of development, thus, lay in agricultural and primary commodities. As economies grow, they accumulate capital, develop skills and generate better technologies and thus the comparative advantage shifts from agricultural and primary products to labour intensive manufactured products, to capital intensive manufactured products and further to skill and technology intensive manufactures. Thus, if analysed in terms of comparative advantage (in static as well as dynamic sense), the nations following the theory of shifting comparative advantage i.e. moving up the ladder of comparative advantage from specialisation in primary products to products which are intensive in capital, skill, technology and knowledge in a gradually progressive fashion would manage to evolve competitiveness in world trade.

Historical experience of the export-led growth of Japan demonstrates that in accordance with the principal of comparative advantage, Japan specialised in the exports of primary goods in the initial stages of development; then in semi-manufactured goods and then into capital intensive manufactures[3]. NIC-4, following Japan, in early 1960s could also achieve higher rates of growth in trade by specialising in exports of primary goods and agricultural products in the initial stages of development, shifting to labour intensive products in late 1960s and 1970s and to capital intensive manufactured products in 1980s. The success of ASEAN-4 in the 1970s in international trade and growth was also based on the same policy. By expanding agricultural exports, ASEAN-4 countries realised greater foreign exchange earnings and improved the linkages between agriculture and other sectors [4].

NIC-4 (except HongKong) and ASEAN-4 having realised that an expanding industrial sector cannot be fuelled for long by resource transfers from a static agricultural sector and the economic potential of agriculture must be mobilised before it can be used, improved their competitiveness by increasing their share of primary exports in the initial stages of development.

The mobilisation of agriculture for general development and trade however would depend, as per the experience of successful Asian countries on:

(a) Macro economic and trade policies which are fairly conducive to expansion of agricultural exports

(b) Protectionist policies aimed at expansion of agricultural exports which gives stress not only on subsidies, but also on an increase in production and quality

(c) The increase in production in turn is achieved by development of adequate infrastructure in the field of agriculture.

The comparative study of 11 semi-indusrializing countries by Bela Balassa [5] reveals that export growth has been particularly rapid in cases where more or less equal incentives have been provided to exports of different sectors and individual export products. Macro economic and trade policies in Korea, Taiwan and ASEAN-4 were fairly conducive to expansion of agricultural exports.

It may not be implied by Bela Balassa's results that only protection will lead to better performance of exports by the agricultural sector of the economy. Unless a sector (industry or agriculture) receiving protection in pushed rapidly to become internally competitive by not only producing more but of quality too, it may not contribute significantly to improvement in competitiveness. The effects of primary exports on industrial development depend to a considerable extent on output relationships. In other words Agriculture could be exploited for general development and trade only if it first produced a surplus which was there for exploitation.

Comparative analysis of growth of agricultural production by countries reveals that agricultural exports were growing fastest in the countries with the highest rates of growth in agricultural production.

Production increase will prove to be effective i.e. cost disadvantage will be reduced only if backed by adequate infrastructure. Infrastructure in the form of ports, railways and roads often represents important inputs for primary exports and their availability may contribute to the development of industrial activities.

The Government is expected to develop the infrastructure by providing various types of economic services for promoting overall developmental climate in the country.

Comparative study of NIC-4, ASEAN-4 and South Asian countries [6] reveals that the share of economic services in total expenditure of Central Government (consolidated accounts) works out much higher for NIC-4, ASEAN-4 and China as compared to South Asian countries.

Thus we may hypothesise that countries whose trade strategy supported composition of trade that not only confirmed to comparative advantage as it existed initially but also changed this composition in accordance with shift in comparative advantage (over time) with major emphasis not only on subsidies but on production and infrastructure improved their competitiveness in world trade.

The competitiveness of exports as per the first part of the hypotheses is to be evolved with development process by changing the composition of trade in accordance with changing or dynamic comparative advantage. For this we have to provide an understanding of the historical forces which give rise to current pattern of comparative advantage and provide an insight into likely direction of change in the future i.e., we have to look into the sources of comparative advantage. Thus before proceeding to the empirical analysis of comparative advantage in exports of South Asian Countries, a review of theories of comparative advantage and empirical models review has been attempted in the following sections to determine the model of export determinant.

3.3 Sources of Comparative Advantage

3.3.1 Introduction

The classical determinants of what goods and services countries will buy and sell in foreign trade is based on the theory of comparative advantage. As formulated by Ricardo and refined by the neoclassicists, the theory identified and arrayed goods in each country according to their unit costs. The precise pattern of specialisation in production and trade depended on comparative costs. The comparative cost principle has profound implications for maximising economic efficiency and welfare.

In order to explain the pattern of international trade, theorists [7] suggest the existence of three broad sets of determinants which govern patterns of specialisation and trade. One set is technological. These determinants can be represented by intercountry variations in relative efficiencies and give rise to comparative advantage through differences in production functions. The Ricardian model of trade provides an example of this line of reasoning. Factor endowments are a second set of determinants. Their effects are highlighted by the factor abundance model of development by E. Heckscher and G. Ohlin (H-O). The third group of determinants is demand related forces. Although these phenomenon have now claimed a place in the literature on trade, there is no equivalent demand oriented theory of international trade since the results it could provide are trivial.

The usual practice, in the absence of a universal model has been to explain comparative advantage in terms of only one set of determinants. The most striking example of this practice is found in the factor abundance approach by H-O. A series of interesting hypotheses and modifications of the simple H-O. theorem examining the importance of various alternative determinants of pattern of trade have been developed over the last forty years. However, during the last few years, there has been considerable scepticism, about the ability of conventional H-O. model to explain the post war growth in world trade specially of trade in manufactures. Neither the extensive trade among the industrial countries, nor the prevalence in this trade of two way exchanges of differentiated products, make much sense in terms of standard theory. As a result many people have concluded that a new framework for analysing trade is needed. The main elements of such a framework--economies of scale, the possibility of product differentiation and imperfect competition have been termed, by Krugman, [8] as New theories of trade among industrial countries.

3.3.2 Ricardian Model

The hypothesis suggested by Ricardian model is that the observed composition of trade can be explained by intercountry variations in comparative costs. Since labour is the key productive factor in the Ricardian model, measures of comparative labour productivity have been designed to serve as a proxy for comparative costs. Given the existence of other productive factors and the fact that, in actuality trade is determined by differences in absolute money prices among countries, the empirical testing of the Ricardian model has foundered on the fact that it clearly ignores many potentially important determinants of comparative advantage.

3.3.3 Heckscher Ohlin Model (H-O)

The basis of trade was originally stated by Ricardo, in terms of the international variations in comparative costs of production. The Ricardian model, however, does not answer the more fundamental question of what determines these variations in costs. H-O model of trade offers a better understanding of the sources of comparative advantage. It is concern for a more complete understanding of the sources of comparative advantage that distinguishes the H-O model from its forebears. The importance of factors of production other than labour were recognised. The relative availability of these factors was seen to be ultimate cause of international trade. This theory attributed to E. Heckscher and G. Ohlin, has received the most attention in the literature.

According to H-O theory, pattern of trade between different countries is determined by the pattern of comparative advantage among them which in turn, (assuming the technology to be the same across countries) is the result of their resource endowment structure. The H-O proposition is that differences in factor proportions are the most important single cause of differences in price structures from country to country [9]. A relatively land abundant country will have a price advantage in respect of those goods which embody much of these land resources and consequently it will be able to export these goods. The basic theorem of the theory developed by Hecksher and Ohlin is the following assertion: "that differential endowments of factors of production between countries is `the most important single factor' in determining comparative cost differences leading to international trade. This coupled with the use of different proportions of the factors of production in different commodities, will lead to a pattern of trade "where each country tends to export commodities using relatively large amounts of its abundant factor" (Heckscher Eli 1949) [10].

The H-O model describes the export and import behaviour of an economy on the basis of its factor endowments and the factor intensities of the products. Notwithstanding the limitations of this theory, several attempts, over the last forty years, have been made in the literature to examine whether the observed trade behaviour confirms to the prescriptions of this theory.

3.3.3.1 Leontief Paradox: In early 1950s, W.W. Leontief [11] made an attempt to verify the validity of the H.O. theorem. Based on the US data of 200 industries for the year 1947, cast in the input output structure with 50 sectors, the study revealed that the manufacture of US exports required a higher proportion of labour to capital than the manufacture of import-competing goods'. Since the US has been traditionally regarded as a country abundantly endowed with capital, these findings clearly conflicted with the expectations of the H-O model. These findings, later termed as `Leontief Paradox' aroused a great deal of controversy in the economic literature. Leontief's methodology, assumptions and figures were questioned and subjected to criticism by numerous people. Leontief himself was not satisfied with his findings. He, in a subsequent article written in 1956 [12], trying to reconcile with findings with the factor proportions approach, concluded that, contrary to earlier expectations the US is indeed a labour abundant country. The quality of US labour makes up for its relatively small quantity and increase productivity to such an extent that labour is relatively abundant in relation to capital. According to the later study by Leontief, with a given quantity of capital, one man year of American Labour is equivalent to, say, three man years of foreign labour. Multiplying, therefore, the actual 1947 labour force of 65 million by three and then finding out capital per worker from this adjusted figure Leontief asserts: "Spread' thrice as thinly as the unadjusted figures suggests - the American capital -supply (per equivalent worker) turns out to be comparatively smaller than that of many other countries. Further Leontief contends that this superiority of American labour vis-à-vis the foreign labour cannot be due simply to large amount of capital which American industry uses per employed worker, but because of the skills, mental attitudes and physical fitness which US labour allegedly possess. He presented a table [13] showing labour inputs in export and import competing industries broken down by skill levels. This table showed that the export industries employed a higher proportion of skilled labourers than the import competing industries. Leontief's latest finding may be reinterpreted as follows:

The international competitiveness of US exports can be ascribed at least in part to the relative abundance of skilled labour. This factor could outweigh the relatively less use by the export industries of the country's other abundant resource - capital.

The Leontief "paradox" has stimulated a series of other interesting hypotheses and modifications of the simple H-O theorem. In particular, a great deal of attention has been devoted to examining the importance of natural resources, human capital, scale economies, technological factors, product cycle, stage of production and consumption patterns as determinants of trade patterns and their composition. The theoretical explanation of these determinants according to Huffbauer [14] (except consumption pattern formulated by Linder), has developed in two directions:

(1) The neo-factor proportions approach and (2) the neo-technological approach.

3.4 The Neo-Factor Proportions Approach

The neo factor proportions approach was directed at attempting to extend the basic two sector H.O. model to include natural resources and to distinguish human capital from physical capital and unskilled labour. Studies of Vanek (1963), Weiser (1968), William (1970) and Baldwin (1971) explain natural resource content and Mitchell, E.J. (1968), Huffbauer (1971), Branson (1971) and Bhagwati (1969) following Leontief, explain the human capital in trade [15].

3.4.1 Human Capital or Skill Explanation of Trade

The inclusion of human capital as a main determinant of the commodity structure of trade was recognised by Leontief and later became a prime focus in empirical implementation of the H-O model. There seems to be little question now that US exports are relatively more human-capital intensive than imports. Similarly, in studying the factor contents of trade in other countries the analysis is enriched when human capital is considered apart from physical capital and raw (unskilled labour. The studies by Bhagwati and Bhardwaj (1970, India) Fels (1972, West Germany), Lovinger (1971, Brazil) Tyler (1972, Brazil) [16] focused narrowly on the role of human capital in determining comparative advantage. The skill explanation of trade is based on the assumptions that the ratio of skilled to unskilled labour is higher in some industries than others, and that interindustry differences in skill requirements are fairly consistent internationally. Thus Donald Keesing argues that international differences in skill endowments, reflecting different rates of investment in labour force, strongly affect the pattern of international trade. Countries, relatively well endowed with skills tend to export skill intensive products if its skill endowment is better than that of its partners.

We may state that the neo-factor proportions approach (directed at attempting to extend the H-O model to include natural resources and to distinguish human capital from physical capital and unskilled labour) enables us to retain the familiar neo classical trade model. The policy guidelines implicit in the H-O theorem - a country should give priority to manufacture of products containing a high proportion of locally abundant factors - are not affected by this interpretation, Factors, however, should be carefully defined in order to avoid possible confusion arising from erroneous classification.

3.5 The New Technology Approach

The neo-technology approach is a more recent and seemingly productive line of approach. The underlying theories - Technological gap, product cycle, economies of scale, stages approach-emphasize dynamic factors in explaining trade patterns.

3.5.1 Stage of Production

Theories of import substitution often presuppose that a stage of production sequence underlines trade. According to import substitution school of thought, the newly developing country will produce for itself and may even export consumer's goods, while the advanced nation will specialise in exporting producer's goods.

3.5.2 Technological Gap and Product Cycle

The product cycle accounts and the technological accounts clearly belong to the same family. Both stress the sequential development of production history. But, while technological gap emphasises time, product cycle emphasises the transition from product differentiation to product standardisation. Hence the two theories merit separate examination.

Technological gap - Technology has been intensively discussed in international trade theory during the last two decades. The discussion focused largely on the impact of change in technology on factor endowment, terms of trade, factor rewards and gains from trade. This approach, however, still retains the basic assumption that all technological changes are known to every producer and that the knowledge of how to produce goods is instantaneously and universally available.

One method of incorporating technology into trade models is to treat it as a capital augmenting factor. Technology under this formulation ceases to be a free factor; it is more expensive in some countries than in others and its intensity in use varies between different industries. This approach has been adopted by G.C. Huffbauer, [17] and H.G. Johnson [18] who suggested that capital should be redefined to include human capital.

An alternative approach is to treat technology explicitly and to introduce an element of time into the trade model. Being first on the market gives a firm substantial advantage over competitors. Some industries have a higher proportion of new goods than others. Some countries have a higher capacity for innovation than others. This approach has been adopted by Posner (1961), Seev Hirsch (1967), G.C. Huffbauer (1970), Vernon (1970) and Klien (1973) [19]. According to Posner (1961) who emphasizes the role of technical progress as a determinant of trade, new products and production processes are constantly being developed and innovating country will, for a time, enjoy a technological superiority over its trading partners. New technologies will eventually be replicated elsewhere but the process of technological advancement is continuously renewed as innovations occur and existing technologies are transferred elsewhere.

Huffbauer concluded that both factor endowments and technological influences were important in explaining the composition of exports. Industries with a high rate of technological progress tend to export a higher proportion of their output than those with a low rate of technological progress. The underlying hypotheses is that the ability to innovate new products and to reduce costs are high when technological progress is high.

Expenditures on Research and Development determine the level and rate of technological progress in an industry. The technology intensive industries tend to be those which have a high share of Research and Development expenditure in the value added. Keesing observed a significantly positive relationship for US industries between Research and Development expenditure as percent of value added and the net exports of an industry.

His findings were confirmed by Gruber, Mehta and Vernon [20]. Gruber and Vernon provided further evidence of the importance of the technological factor (defining technology intensive industries as those which employed scientists and engineers in excess of six percent of their total work force, as measured by US data) by examining its role in a world trade matrix. Maurice Girgis analysis also supported the hypothesis that trade patterns in manufactured goods could be attributed to international differences in levels of technological achievements.

3.5.3 Product Cycle Model

The product cycle model of trade, developed by Hirsch (1967), [21] is based on the hypotheses that factor intensities as well as technical and marketing characteristics of different product groups vary in a systematic way with the maturity of a product. According to Hirsch, the advanced countries tend to have a comparative advantage in "new" and "growth" products because of their relative abundance of scientific and engineering know-how, capital management and external economies, namely factors important in the manufacture of these products, i.e., products with a stable and well established production process. "Mature" products are less demanding in their requirements of inputs that are relatively scarce in LDCs-Scientific know-how and capital - but are intensive in the use of the relatively abundant factor namely unskilled labour (Textiles, e.g.) Hence industrialising countries should specialise in the production and export of industries which have reached a matured stage in developed economies.

Another version of the model (Vernon, 1979) [22] proposes that the development of new products is contingent on the nature of demand in the richer countries. Production and exports - will first occur in these countries and will only later be transferred to lower cost locations.

Partial tests of the product cycle model have been carried out by Wells (1969) and Adler (1970) [23]. Wells formulations were unfortunately vague and empiricism was so casual that his results do not seem well grounded. Similarly, Adler's assumption about the relations involving price elasticities was shown to be valid.

3.5.4 Availability Hypotheses

Availability hypotheses, developed by Kravis, explains a country's trade composition, at a given point of time, in novel way. According to this, hypotheses, a country imports goods which happen to be non-available at home and exports those that are available. According to Kravis, "it is the elasticity of supply abroad and its inelasticity at home that gives rise to this trade, not the relative capital or labour requirements of the products". Natural resources, innovation (technological progress) and product differentiation are the important determinants of availability. The product cycle thesis associated with Vernon and others is in fact the same as Kravis' availability thesis. National product differentiations - in the sense that due to preferences of buyers the products of a particular nation succeed in securing a favoured position in a country - can also be seen as a basis for the operation of availability factor in international trade.

3.5.5 Preference Similarity

S.B.Linder [24] offers a different story of determinants of trade. According to Linder, "trade in primary products may be due to differences in natural resource endowments among countries and therefore the H-O theorem provides a reasonably good explanation of trade in these products. Trade in manufactures, on the other hand, is due to structural similarities between countries. It is possible that the factor proportion theorem has gained such wide acceptance only because a pattern of trade in primary products dictated by differences in natural resource endowments, is so plausible".

Linder's central hypothesis is that the intensity of trade in manufactures between two countries is greater, the greater the similarity in the demand patterns of the two countries. The pattern of demand is further assumed to be similar in countries with comparable levels of per capita income. In other words, the volume of trade in manufactures will be more intensive between countries with similar per capita income than between countries with dissimilar incomes. This theory is based on two premises:

(a) It presumes that existence of home demand is a necessary condition for a product to be qualified as export.

(b) As a corollary to the above, the output composition of a country is supposed to adjust itself to the pattern of its internal demand, as the relative prices of the country's product will then tend to be low.

For a product to be consumed (or invested) in a country, there must be a demand for the product at the ruling world market prices. It is representative demand that is necessary for a good to be potential export product.

Linder gives two reasons for this hypothesis:

(a) It is unlikely that an entrepreneur will even think of satisfying a need that does not exist at home;

(b) Even if foreign need was seen, the basically correct product to fill it might not be conceived of, it is still improbable that the product could be finally adapted to unfamiliar foreign conditions without prohibitive costs being incurred. The exceptions; according to Linder, to the home demand prerequisite are (i) it is easy to become aware of foreign demand in spite of the non-existence of home demand for the product; (ii) the product as such is available without inventive effort; and (iii) little or no product development work is needed. He thought primary product industries would fall into this category and the factor proportions account of trade to be applicable.

A review of the two approaches reveals that the interpretation of the Neo-factor proportion approach does not affect the policy guidelines implicit in the H-O theorem, a country should give priority to the manufacture of products containing a high proportion of locally abundant factors. Factors, however, should be carefully defined in order to avoid possible confusion arising from erroneous classification. In other words, "the neo-factor proportions enables us to retain the structure and methodology of the familiar and elegant neo-classical trade models, though it may require the introduction of one or two additional variables"(Hirsch, 1974)[25]

The neo-technology approach emphasising dynamic factors such as technological gap, product cycle, economies of scale, in explaining trade patterns is a more recent and productive line of approach. Though these factors are important and they need to be distinguished more carefully, the factor endowment model has yet to be integrated systematically with an endogenous mechanism of technological change and diffusion [26]. A noteworthy effort to incorporate neo-technology explanation of trade with the neo-factor proportions was made by Huffbauer in 1970 [27].

Huffbauer sought to explain the characteristics embodied in 1965 exports for 24 developed and less developed countries, using 3 digit SITC trade data for manufactures and assuming that US coefficients could be applied to all other countries, he calculated export characteristics for each country that were designed to reflect (i) capital per man, skill ratios, and wage differentials in accordance with the factor endowments theory; (ii) the "first trade date", to represent the introduction of new products according to the theory of the technological gap; (iii) Product differentiation, to represent a stage of specialisation in the product cycle; (iv) scale economies arising from specialisation for the home market; and (v) the ratio of consumer goods to total industry sales, to capture the stage of economic development. These national export characteristics were then related by rank correlation methods to various national attributes: fixed capital per man in manufacturing and skilled employees as a percentage of the labour force to approximate factor endowments; gross domestic product per capita to reflect technological sophistication; and total manufacturing output to represent the possible realisation of scale economies and the stage of development in production. Despite the crudity of his measures Huffbauer found that the various explanations did relatively well and concluded that both factor endowments and technological influences were important in explaining the composition of exports.

Leamer's latest work (1984) [28] may be considered to be an improvement over Huffbauer's analysis.

The simple linear model by Leamer explained a large amount of variability of net exports across countries and it also identified the sources of comparative advantage. As per Leamer's results "unskilled labour and land are sources of comparative advantage in the agricultural products and natural resources are sources of comparative advantage in natural resource products. As per manufactured products, in 1958 success in exporting manufactured products was linked to abundance of knowledge capital embodied in the professional/technical work forcc. In 1975, however, knowledge capital continued to be the most important source of comparative advantage in chemicals but physical capital and skilled and unskilled labour were the sources of comparative advantage in machinery and labour intensive and capital intensive manufactures" [29].

There are several novelties in Leamer's work as compared to Huffbauer's.

Firstly, Huffbauer's study was confined to manufacturing sector only whereas Leamer explained the structure of trade for all the sectors of an economy.

Secondly, Huffbauer assumed that US coefficients could be applied to all other countries whereas Leamer did not have to impose US coefficients on other countries.

Thirdly, Leamer's work offered an explicit procedure for selecting `best' theory by the multiple testing of competing theories in contrast to Huffbauer's procedure of testing theories one by one.

Finally, Huffbauer searched for a single theory for all commodities that could vary from country to country whereas Leamer's work profiled each country's characteristics on the basis of the commodity cross section results by allowing the best theory to vary with commodity class while assuming it constant across countries.

Leamer's model is a static model of comparative advantage. Although Leamer mentions about change in accumulation of capital in the developing countries from 1958 to 1974, his model does not incorporate the dynamic characteristics of the pattern of international trade or comparative advantage. In a rapidly changing world of capital accumulation and technological progress, the dynamic nature of comparative advantage needs to be stressed.

3.6 New Theories of Trade Among Industrial Countries

There has been considerable scepticism during the last few years, about the ability of conventional H-O model to explain the post-war growth in world trade specially of trade in manufacturers. Many economists have proposed that a new framework for analysing trade is needed. The main elements of such a framework -economies of scale, the possibility of product differentiation and imperfect competition- have been termed, by Krugman, as New theories of trade.

The conventional forces of comparative advantage operate on groups of product (industries) and thus give rise to interindustry specialisation and trade. Several recent empirical studies of trade suggest that inter-industry specialisation and trade, which reflect the conventional forces of comparative advantage, are also accompanied by intra-industry specialisation, which reflects scale economies and consumer's taste for a diversity of products. Much of the world's trade in manufactures is trade between industrial countries with similar relative factor endowments; much of the trade between these countries involves two-way exchanges of goods produced with similar factor proportions (intra-industry trade). Furthermore most manufacturing industries are characterised by at least some degree of increasing returns (specially if we include dynamic scale economies associated with Research and Development and the learning curve). Most manufacturing industries are also imperfectly competitive to at least some extent. Thus conventional model (H-O) which assumes constant Returns, exogenous technology and perfect competition cannot give adequate guidance for trade policy in these industries.

In response to this, many economists [30] have proposed alternatives to conventional trade theory. It has been widely recognised that economies of scale provide an alternative to differences in technology or factor endowments as an explanation of international specialisation and trade. Balassa (1967) and Kravis (1971) [31] have argued that scale economies play a crucial role in explaining the post-war growth in trade among the industrial countries. Assuming scale economies to be internal to firms, with the market structure that emerges being one of chamberlinian monopolistic competition, the model shows that trade need not to be a result of international differences in technology or factor endowments. Instead, trade may simply be a way of extending the market and allowing exploitation of scale economies, with the effects of trade being similar to those of labour force growth and regional agglomeration.

Thus, according to new theories of international trade, scale economies play a crucial role in explaining the post-war growth in world trade among the industrial countries - perhaps more important than differences in factor endowments.

3.7 Theoretical Validity of The H-O Model

The H-O Model has conventionally been used to explain the pattern of international trade. According to this theory (assuming the technology to be the same across countries), comparative advantage is the result of the resource endowment structure of different economies. The model describes the export and import behaviour of an economy on the basis of its factor endowments and the factor intensities of the products. A series of modifications of the simple H-O theorem, examining the importance of natural resources, human capital, scale economies, technological factors, product cycle, stage of production and consumption patterns have been developed over the last forty years. The theoretical explanation of these determinants has developed into three directions :

(a) The neo-factor proportions approach

(b) The neo-technological approach, and

(c) New Theories of Trade

A detailed review of the first two approaches reveals that :

The neo-factor proportions enables us to retain the structure and methodology of the neo-classical trade models, though it may require the introduction of one or two additional variables. Factors, however, should be carefully defined in order to avoid possible confusion arising from enormous classification.

The neo-technology approach emphasises dynamic factors such as technological gap, product cycle, economies of scale. Though these factors are important and they need to be distinguished more carefully, the factor endowment model has yet to be integrated systematically with an endogenous mechanism of technological change and diffusion. A noteworthy effort to incorporate neo-technology explanation of trade with neo-factor proportions was made by Huffbauer (1970). Leamer's work (1984) may be considered to be an improvement over Huffbauer's analysis. Leamer's model, however, is a static model and does not incorporate the dynamic characteristics of comparative advantage. The need is to improve our understanding of the processes whereby technology and other forces determine comparative advantage at a point in time and alter its structure overtime. To accomplish this, these various forces must be treated endogeneously in our theoretical model. The factor endowment model has yet to be integrated systematically with an endogenous mechanism of technological change and diffusion. According to Stern, due to conceptual inadequancies in recent empirical research on technological and related factors, explanations of trade were often difficult to interpret and compare. Thus it may not be possible to integrate factor endowment model systematically with an endogenous mechanism of technological change and diffusion at a point in time.

We may contend that all the dynamic factors are indirect source of comparative advantage and thus govern the future course of comparative advantage over time by changing the factor endowments first and then the pattern of trade. For example improvement in technology will lead to change in factor endowment of an economy. This changed factor endowment will lead to change in the pattern of trade. Thus, though it may not be possible to integrate factor endowment model systematically with an endogenous mechanism of technological change and diffusion at a point in time as suggested by Stern, it may be feasible to do so by testing empirically the factor proportions model over different points of time in an economy. What we require is to determine the composition of exports as per different factor endowments at different points of time as the factor endowments change with changes in technological factors.

Every economy, initially is an underdeveloped economy characterised by low level of technology and skill, abundant labour, shortage of capital and small sized domestic market. Its pattern of trade is determined by natural endowments i.e. land, labour and capital. Its composition of exports will comprise more of raw material products, agricultural products or labour intensive manufactures depending upon the relative endowment of raw materials, land or labour and capital in the economy. This resource endowment of the economy changes with technological improvements. In turn, technological improvements lead to the relative endowment of capital. With this increased accumulation of capital as compared to natural resources and labour, the pattern of specialisation changes. Economy's share of exports of light manufactured products increases as compared to raw material and agricultural products. Once the process of technological improvement changes the resource endowment of an economy the process of development goes on. Increased capital accumulation leads in turn to better skill and technological advancements which changes the pattern of specialisation further from light manufactured products to skill and technology intensive manufactured products. Thus factor endowment proposition's structure is retained if observed as a static one at different dynamic points i.e. at different points of time as per level of technological achievement. Bela Balassa's theory of stages approach to comparative advantage strengthens the above view point.

Thus, we may state that neo-factor propositions approach as well as neo-technological approach to comparative advantage enables us to retain the structure and methodology of factor proportions theorem. The policy guidelines implicit in the H-O theorem are not affected by these interpretations. The Neo factor proportions require the introduction of one or two additional variables and the Neo-technology approach enables us to explain factor proportions theorem at different points of time as per different levels of technological advancement and hence level of development achieved. In other words, the sources of comparative advantage lay in differences in factor endowments. The factors, however are to be defined clearly. Factor endowment is the result of (1) natural endowment and (2) past economic, political and social process. We may state that endowment of land, labour and part of physical capital is the result of natural endowments whereas that of capital (human, physical) skill, technology are the result of past, economic, political and social processes. The importance (weightage) of different factors may vary from country to country depending on level of development) and from sector to sector. Differences in natural endowments (land, labour and natural resources) may be given more weightage as sources of comparative advantage in raw materials and agricultural products, capital (physical, human), technology, scale economies, research and development may be given more weightage in manufactured products. Secondly, the distribution of resources across countries is changing overtime but the pace of change is different for each resource. Patterns of specialisation and trade should be altered as a result of these shifts, i.e., attempt should be made to evaluate the dynamic comparative advantage as compared to static one.

The review of the third approach i.e. New Theories of trade reveals that, there has been considerable scepticism during the last few years, about the ability of conventional H-O model to explain the post-war growth in world trade specially of trade in manufacturers. It has been widely recognised that economies of scale provide an alternative to differences in technology or factor endowments as an explanation of international specialisation and trade. The role of scale economies, as per new theories of trade, is considered to be more important than the differences in factor endowments in explaining the post war growth in world trade among the industrial countries.

The strength of two sets of determinants (factor endowments and scale economies), however, would vary depending on nature of trade (inter-industry or intra-industry). Helpman and Krugman (1985, p.262) [32] show that the pattern of trade is shaped by the underlying exchange of factors but with an overlay of additional specialisation to realise economies of scale: the pattern of inter-industry trade is explained by the conventional H-O model, while the overlay of additional specialisation is intra-industry in scope and its extent is dependent upon country similarities. "Countries with similar factor endowments will engage in intra-industry trade, while countries with different endowments will engage in H-O trade" (Krugman, 1990) [33].

The analysis of international distribution of factor endowments by Leamer (1984) [34] and Ballance (1990) [35] reveal substantial differences between the two major country groups, i.e., Developed Market Economies (DME) and Developing Economies (DE). The study by Ballance (1990) further reveals that the differences in the endowment patterns of the developing countries are as interesting as the distinctions between two country groups (DME and DE). DMEs are relatively homogenous in terms of their basic economic characteristics while DE's are not. Secondly, scale economies and product differentiation may exert a positive influence on intra-industry-trade in industries where concentration is significantly and positively influenced by the extent of economies of scale while the intra-industry trade of DEs reveals no similar evidence.

Thus, as the diversity of factor endowments is great, a strong H-O component (with modifications) should be present among the determinants of the commodity composition of trade, specially for developing economies of South Asia.

3.8 Conclusion

A review of the theories of comparative advantage and the empirical models review reveals that the H-O Model has conventionally been used to explain the pattern of international trade. According to this theory (assuming the technology to be the same across countries), comparative advantage is the result of the resource endowment structure of different economies. A series of interesting hypothesis and modifications of the simple H-O theorem examining the importance of various alternate determinants of pattern of trade have been developed over the last forty years. A detailed review of these modifications reveal the following: First, sources of comparative advantage lay in differences in factor endowment. The weight carried by different factors may, however, vary from country to country and from sector to sector. Differences in natural endowment (land, labour and natural resources) have been taken to be as sources of comparative advantage in case of raw-materials and agricultural products. Capital (physical and human), technology, scale economies, research and development have similarly been given high weight in manufactured products. Second, in a world of rapidly changing technology one needs to focus on the dynamics of comparative advantage as compared to the static one. Third, the role of scale economies in determining pattern of trade would depend on the nature of trade (inter-industry or intra-industry) which in turn is dependent upon country similarities. "Countries with similar factor endowment will engage in intra-industry trade, while countries with different endowment will engage in H-O trade (Krugman, 1990). As there as substantial differences in the endowment pattern of developing economies of South Asia, a strong H-O component should be present among the determinants of the commodity composition of trade in these economies. The comparative advantage of trade in agricultural products of South Asian countries, thus, is examined (in static as well as in dynamic sense by practical application of the commodity version of H-O model.

The familiar neo-classical version of factor proportions explained the pattern of trade for the economy as a whole i.e. considered all the sectors - agriculture, industry etc. The later theories, except a few (Leamer, Linder) considered sources of comparative advantage in manufacturing sector only. The reason may be that most of the studies were for developed economies (mainly USA) who were at that stage of development where primary exports weightage becomes quite low. South Asian economies being underdeveloped economies characterised by abundance of labour, scarcity of capital, technology etc. our analysis will consider not only the manufacturing sector but all the sectors in these economies to judge whether their pattern of trade has been as per the resource endowment structure as reflected in the comparative cost advantages.

Notes

1. Report of the President's Commission on Industrial Competitiveness - U.S.A. (1982).
2. Banerjee, R, (1975): - Export of Manufactures from India.
3. Tripathy, B.N. (1985) Exports and Economic Growth.
4. James, W.E., Naya, Seiji & Meier, G.M. (1987): Asian Development.

5. Balassa, Bela (1982) - Development Strategies and Economic Performance: A Comparative Analysis of Eleven Semi-Industrial Economies.

6. Please See Chapter 6.

7. For details please see:

(i) Stern R.M. (1980): Testing Trade Theories. and (ii) Ballance, Robert,H. (1987) : International Industry and Business.

8. Krugman, Paul R. (1981) : Interindustry Specialisation and the Gains from Trade.

9. H-O theory of comparative advantage is available in two versions. (a) The factor content version, (b) The commodity version. The factor content version depicts trade as an implicit exchange of the factors of production embodied in traded goods. The commodity version suggests that a country will tend to export goods which require comparatively large amounts of those factors of production which are relatively abundant in the country. Thus, a country which is rich in capital would be expected to produce and export goods which require relatively large amounts of capital.

10. Heckscher, Eli (1949) ; "The Effects of Foreign Trade on 'the Distribution of Income" Readings in the Theory of International Trade (Philadelphia, 1949, p; 285.

11. Leontief, W. (1953) : Domestic Production and Foreign Trade : the American Capital Position Reexamined.

12. Leontief, W. (1956) : Factor Proportions and Structure of Trade: Further Theoretical and Empirical analysis.

13. For details of the Table see, Leontief, Factor Proportions and the Structure of American Trade. Further Theoretical and Empirical Analysis, Review of Economics and Statistics November 1956.

14. Huffbauer, G.C. (1971) : The Impact of National Characteristics and Technology on the Commodity Composition of Trade in Manufactured Goods.

15. For details please see Bibliography.

16. For details please see Bibliography.

17. Huffbauer, G.C. (1970) : Impact of National Characteristics and Technology on the Commodity Composition of Trade in Manufactured Goods.

18. Johnson, H.G. (1968) ; Alternative Maximization Policies for Developing Countries Exports of Primary Industry.

19. For details please see Bibliography.

20. William H.Gruber, Dileep Mehta, Raymond Vernon (1967) : "The R and D Factor in Industrial Trade and International Investment of United States Industries".

21. Hirsch, S. (1967) : Location of Industry and International Competitiveness.

22. Vernon (1979) : An Anatomy in the Theory of Comparative Advantage.

23. Adler, F.M. (1970) : The Relationship between the Income and Price Elasticities of Demand for United States Exports.

24. Linder, Staffan, Bruenstan (1961) : Essay on Trade and Transformation.

25. Hirsch; S. (1974) : Capital or Technology.

26. Stern, R.M. (1980) : Testing Trade Theories.

27. Huffbauer, G.C. (1970): The Impact of National Diaracterstics and Technology on the Commodity Composition of Trade in Manufactured Goods.

28. Leamer, Edward, E. (1984) : Sources of Comparative Advantage.

29. Leamer : Ibid.

30. Balassa (1967), Grubel (1970), Krugman (1980, 1981, 1986,1990) Helpman and Krugman (1985,1989). Please see Bibliography for details.

31. Kravis I.B. (1970) : Comment on Huffbauer and Gruber and Vernon, in Vernon ed. (1970).

32. Krugman, and Helpman E (1985): Market Structure and Foreign Trade.

33. Krugman, Paul. R. (1990): Increasing Returns and Economic Geography.

34. Leamer (1984): Sources of Comparative Advantage.

35. Ballance, R. and Melmut F. (1990): Competing in a Global Economy.

Chapter 4

Resource Endowment Structure and Comparative Advantage

4.1 Introduction

South Asian economies, at present, confront a crisis in their international competitive position. The staggering trade deficits of the past several years and the borrowing that has financed them cannot be sustained over the long run. Policy makers will have to grapple with ways to improve export performance.

The analysis of export performance of South Asian countries as compared to NIC-4 and ASEAN-4 (Chapter 2) revealed that the root cause of poor performance of exports in South Asian countries is lack of international competitiveness. "A country's ability to compete effectively with other sources of supply in the world market is perhaps the most crucial determinant of exports and export growth" (Deepak Nayyar, 1975) [1]. Thus enhancing competitiveness may be termed as one of the major objectives of trade strategy for South Asian countries. Competitiveness of exports may depend on whether the composition of exports, as suggested by trade theories is determined by comparative advantage or not. The comparative (cost) advantage has two useful application - (a) it serves as one of the fundamental concepts in descriptive (or positive) theories by providing a basic explanation of the international pattern of specialisation in production and trade, and (b) it plays an important role in prescriptive (or normative) economics by providing guidelines for government policies on resource allocation and trade. In the positive role, trade theories are able to tell us why, after all, the composition of exports differ from one country to another.

It was hypothesised (Chapter 3) that countries whose trade strategy supported pattern of trade that not only conformed to comparative advantage as it existed initially but also changed the strategy in accordance with shifts in comparative advantage (over time) improved their competitiveness in world markets. Further, the review of trade theories, attempted in Chapter 3 reveals that the pattern of trade between different countries is determined by the pattern of comparative advantage among them which in turn, according to H-O theory is the result of the nature of their resource endowment structure. The sources of comparative advantage, in other words, lay in difference in factors endowments as explained by neo-classical theory of factor endowment. However, in practice, the trade flows constitute the net results of (i) the ideal comparative advantage based on resource endowment structures and (ii) the distortions introduced by trade policies such as controls, quota restrictions, tariffs, licensing system, subsidies and export promotion incentives etc.

As noted earlier, ideal comparative advantage for developing South Asian economies, characterised by abundance of labour, scarcity of capital, relatively low level of technology and heavy dependence on natural endowments, would seem to lie in agricultural products (which depend on natural endowments) and labour intensive manufactures (which entail normally low skills and simple technology). However, most of these countries (specially India, Pakistan and Bangladesh) from beginning of 1970s had an active policy of promoting the exports of new manufactures and neglecting the traditional exports. According to James, Naya and Meier (1987) [2], "In most South Asian countries, policy produced price distortions and direct government interventions adversely affected industrial development. These policies aimed at industrial self-sufficiency (specially in basic industries, e.g. chemicals, metal products, machinery) under close government supervision, and largely disregarded the countries' factor endowments to industrial development (such as insufficient infrastructure, energy shortages and lack of organisational and managerial skills). A whole range of subsidies and other incentives were offered to exports of manufactures." The structure of their global merchandise, imports and exports (Table 4.1) exhibits structural changes away from exports of primary goods to manufactures and also in imports - a move towards capital goods and intermediate products.In the light of the hypothesis and theoretical analysis of Chapter 3, an attempt is made in the present chapter to verify empirically whether this pattern of trade of South Asian economies conforms to the structure of comparative advantage as reflected in their resource endowment structure (in static as well as in dynamic sense). However, due to vastness of the subject, analysis is divided into two chapters. The present chapter, i.e. Chapter 4 attempts to examine the proposition that the comparative advantage for South Asian countries lies in the exports of agricultural products and labour intensive manufactures. The proposition is examined by the empirical verification of H-O model of trade. Having established that the comparative advantage for South Asian countries lies in exports of agricultural products and labour intensive manufactures, we analyse, in the next chapter, the composition of exports of these countries vis-à-vis comparative advantage to see whether the composition of exports of these countries has been according to comparative advantage or not.

4.2 Empirical Verification of H-O Model of Trade: South Asia

The analysis takes into account both the static and dynamic nature of comparative advantage. The static nature of export compatibility is analysed by an application of the commodity version proposition of the Heckscher-Ohlin (H-O) theory [3]. The dynamic nature of compatibility of South Asian countries' exports is analysed by testing the Bela Balassa's theory of "The stages

Table 4.1: Structure of Merchandise Exports/Imports of South Asian Countries 1965 and 1988

Country	Year	Exports			Imports		
		Foods, Minerals Metals	Other Primary Commodities	Total Manufactures	Foods, Minerals Metals	Other Primary commodities	Total manufactures
India	1965	10	41	49	27	14	59
	1988	9	18	73	18	9	73
Pakistan	1965	2	62	36	23	5	72
	1988	1	30	69	33	13	53
Bangladesh	1965	NA	NA	NA	NA	NA	NA
	1988	2	29	69	33	9	58
Sri Lanka	1965	NA	NA	NA	49	4	46
	1988	7	50	43	34	3	64

Source: Regional Economic Cooperation Between SAARC, ASEAN and Selected Other Countries of Asia Pacific Region - Charan D. Wadhwa.

approach to comparative advantage" (Balassa, 1965) [4]. Balassa uses the commodity composition of trade as the reference pattern to gauge anticipated changes in comparative advantage on the basis of changes in country's factor endowment. In other words, the dynamic comparative advantage is observed by testing empirically the factor proportions model as a static one at different dynamic points. As the concept of comparative advantage involves intercountry comparisons, we attempt a comparison between South Asian countries, and some of the successful developing Asian countries, NIC-4 and ASEAN-4.

A practical application of the proposition requires detailed knowledge of factor endowments and the intensity of factor usage in various industries, as well as the trade composition of the country concerned. In other words, a three way relationship between factor abundance, factor intensity and trade is stipulated for each country and its trading partners [5].

4.2.1 Methodology

Various methods [6] can be employed for a practical application of the commodity version of H-O model which requires a detailed knowledge of factor endowments, factor intensity and trade composition of the country concerned. The best one would involve a rigorous econometric analysis carried out on the basis of an appropriately specialised model of the determination of comparative advantage. Such an analysis, however, would be difficult to undertake at least for three reasons:.[7].

(a). The complexity of trade theory considerably complicates the specification of such a model.

(b). The data required for the analysis would have to cover a wide spectrum of resources and exhibit a design that allowed for sufficiently precise statistical inference.

(c). It requires a sequence of incredible assumptions.

In view of the above difficulties Huffbauer [8] proved the validity of the trade model by demonstrating reasonable matching between a country's attributes and the goods exported. S. Hirsch [9] in his paper, gave additional demonstration that matching between country and industry attributes does indeed yield comparative advantage. Leamer (1984) [10] used different media words, tables, graphs, simple correlations between trade data and the resource data and also the econometric analysis of multiple correlations to clearly indicate the composition of trade and relative abundance of resources in analysing the sources of comparative advantage. A U.N. study [11] conducted in 1986, confined itself to descriptive statistical methods to relate the distribution of resources among countries to the pattern of comparative advantage. The concept of "factor-orientation of Revealed Comparative Advantage" (RCA) was introduced in this study.

The methodology adopted in the present study is a combination of regressions or simple correlation methods by Leamer [12] and interaction between factor endowment and factor intensity by Hirsch (1974) [13]. According to Hirsch, "comparative advantage is the outcome of interaction between two variables: factor endowment which is a country characteristic and factor intensity which is an industry or product characteristic. The procedure followed is:

(a). To estimate the factor endowment of South Asian countries:

(i) Absolute factor endowment using Resource Abundance Ratio [14] as an indicator of factor abundance.

(ii) Relative factor endowment in physical terms using ratios of (1) gross fixed capital formation to unskilled labour and land, (2) skilled labour to unskilled labour and (3) land to labour as proxy variables [15] to compare the relative endowment of capital and skilled labour to unskilled labour and land for South Asian countries as compared to NIC-4 and ASEAN-4.

(b). To establish factor intensities of different sectors i.e. agricultural, labour intensive manufactures and capital intensive manufactures.

(c). The interaction between the above two i.e. factor endowment and factor intensity enables us to determine ideal comparative advantage (in static sense) for South Asian countries.

(d). The dynamic comparative advantage is observed by considering the period between 1970 and 1985 and estimating the relative changes in factor endowment of South Asian countries, NIC-4 and ASEAN-4.

Before we proceed to empirical analysis, a brief review of the concepts involved is given below.

4.2.2 Resource or Factor Endowment

Factor endowment, for an economy is given, at least in the short run. It is the result of natural endowment as well as past economic political and social processes [16].

Factor endowment for an economy can be measured in two ways: (a) absolute factor endowment and (b) relative factor endowment. The indicator of absolute factor endowment, in our analysis is Leamer's resource abundance ratio i.e.

$(V_i/V_{wi}) / (Y/Y_w)$ [17] where

V_i - denotes endowment of resource i in country X,
V_{wi} - world's endowment of resource i,
Y - GNP of country X and
Y_w - world GNP.

The country is abundantly supplied in factor i compared with other resources if an endowment share (V_i/V_{wi}) exceeds GNP share (Y/Y_w) i.e.

$(V_i/V_{wi}) > (Y/Y_w)$ or $(V_i/V_{wi}) / (Y/Y_w) > 1$

4.2.3 Relative Factor Endowment

"Measurement of relative factor endowment and their trend over time is replete with difficulties. Since physical quantities at the aggregate level can not be meaningfully conceived (or measured), the value of gross fixed capital formation (GFCF) in constant prices would be used to measure the factor endowment in reproducible tangible capital" [18].

Hence for present analysis,

(a) GFCF to unskilled labour ratio and GFCF to land ratios are considered proxy variables to compare the relative endowment of capital to labour and land, and

(b) Skilled labour to unskilled labour ratios and skilled labour to land ratios are considered proxy variables to compare the relative endowment of skilled labour to unskilled labour.

(c) Land to labour ratios are considered the proxy variables to compare the relative endowment of land to labour.

4.2.4 Factor Intensity

Factor intensity is a technological parameter which varies from one industry to another and change over time as a result of technological improvements in production processes (Ballance, 1987) [19]. Commodity X is said to be the relatively capital intensive commodity if $(aK_1/aL_1) > (aK_2/aL_2)$ Where (aK_1, aL_1) and (aK_2, aL_2) are the amount of capital and labour required to produce a unit value of X_1 and X_2 respectively.

The concept of factor intensity may be defined to include inputs directly entering into the production process. For example, the measurement of direct factor intensities might take account of the amounts of labour and capital actually employed in the industry. In this way, labour and capital embodied in the intermediate inputs used by the industry would be ignored. An alternative approach is to construct a measure of total factor intensity which would incorporate not only direct factor usage but also the labour and capital that enter with the production process indirectly via intermediate inputs. With certain qualifications there is general agreement that total factor intensities are the appropriate choice for studies intended to explain patterns of trade, while direct factor intensities are more relevant for an analysis of international pattern of production. Thus recommendation depends mainly on the extent to which intermediate inputs are traded. If they are widely available, the relevant factor requirements are only those which apply to each particular stage of production.

Factor intensity may be measured in two ways: one possibility is to express the measure in terms of a stock--for instance, the total amount of capital or labour available in the industry. Examples of stock measures would be the accumulated physical capital per employee or an aggregation of man years used in the industry. Alternatively, factor intensities can be pictured as a flow expressed in terms of the payments to labour and capital for services rendered during the accounting period. Both measures are proxies involving certain drawbacks and have been debated extensively in the relevant literature. For instance, estimates of capital stock are derived from data on capital assets (building, machinery and equipment) and are expressed at different price levels depending on their date of acquisition. Differences in rates of depreciation and obsolescence, as well as the use of historical rather than replacement values for physical capital, represent disadvantages of the stock measure.

With regard to flow measure of factor intensity, for the actual measurement of factor intensity, the income or value added generated in any productive activity is attributed to the use of physical capital (e.g. machines and equipment) and to labour. Human participation in the production process represents the contribution of rudimentary labour and application of special skills. Accordingly, the share of wages and salaries in value added can be pictured as consisting of two components that is, the compensation for unskilled labour and the returns to skills. Furthermore, the difference in pay of an unskilled labourer (one with only minimal education and training) and that of a skilled labourer represents the returns paid for the skills acquired through education and training. To the extent that differences in wages and salaries reflect differences in productive skills, an industry offering relatively high average wages and salaries per employee may be regarded as one which is intensive in its human capital requirements. Similarly, in the absence of excessive profits the relative level of the non wage and salary component of value added per employee in an industry reflects the intensity of the use of physical capital. Like the use of stock measure, estimates of factor intensities that are based on the contribution of each input to total value added have several shortcomings.

The discussion clearly implies that the empiricist's tools for determining factor intensities are limited. At best, only approximations can be expected. Despite the drawbacks, however, many studies have concluded that such measures provide a useful basis for examining international trends in the use of labour and capital.

In a study by Leamer (1984), the estimates of factor intensities were based on a concept of direct factor usage and were expressed as stock, rather than a flow. Three factors i.e. land, labour and capital, were used in the analysis. The procedure adopted was to establish a typology of industry through the factor orientation indices [20]. This typology of industries gave us factor intensity of a product or a sector in the economy. As establishment of this typology involves a large amount of trade, endowment and GNP data across countries, it is beyond the scope of our study to arrive at this typology. In a number of empirical tests of the H-O theory (Huffbauer and Balassa) [21] the highly simplifying assumption is made (at least implicitly) that factor intensities of commodities are identical across countries. Assuming the same, the typology of industries given by Leamer for different sectors is used in our analysis.

4.2.5 Interaction between Endowment and Intensity

Comparative advantage is the outcome of the interaction between two variables factor endowment, which is a country characteristic and factor intensity, which is an industry or product characteristic [22]. Comparative advantage is empirically taken to be high when

Table 4.2: Impact on CA by Interaction between Endowment and Intensity

Country Endowment	Product Intensity	Comparative Advantage
HIgh	High	High
High	Low	Low
Low	High	Low
Low	Low	High

the values of the factor intensity and the factor endowment go together. Comparative advantage is taken to be low in other cases. The outcome of these interactions, assuming for expository purposes that endowment, intensity and comparative advantage can have only two values - high and low - are shown in Table 4.2.

4.3 Comparative Advantage: South Asia (Empirical Results)

4.3.1 Absolute Factor Endowment: South Asia

Leamer's Resource Abundance Ratio is used as an indicator of resource abundance for India, Pakistan, Bangladesh, Sri Lanka and Nepal. For all these countries, (except for Bangladesh for land and Sri Lanka for land in 1970 and 1975), Absolute Resource Abundance Ratio for land and labour was positive and greater than one for the year 1970, 1975, 1980, 1985 and 1989) (Table 4.3).

4.3.2 Relative Factor Endowment: South Asia

The ratio of GFCF to unskilled labour, GFCF to land, land to labour and skilled labour to unskilled labour are considered proxy variables to compare the relative endowment of (i) capital to unskilled labour and land, (ii) land to labour and (iii) skilled labour to unskilled labour for South Asian countries, Korea, Singapore and ASEAN - 4 countries (Table 4.4).

Relative endowment of capital and skilled labour has been found to be quite low in South Asia as compared to Korea, Singapore, Malaysia, the Philippines and Thailand in each of the years 1970, 1975, 1980 and 1985. GFCF/unskilled labour ratio was lowest in South Asia (except Sri Lanka) as compared to all

Table 4.3: Absolute Resource Abundance Ratio : Land and Labour, South Asia.

Country	Year	Y/Yw	Vi_{land}/Vw_{land}	$Vi_{lab.}/Vw_{lab}$	Land Abundance Ratio	Labour Abaundance Ratio
India	1970	0.01809	0.02274	0.14019	1.25670	7.74860
	1975	0.01511	0.02274	0.13808	1.50498	9.13898
	1980	0.01486	0.02273	0.13555	1.52660	9.12093
	1985	0.01677	0.02273	0.13551	1.35571	8.08125
	1989	0.01383	0.02274	0.13616	1.68642	10.09862
Pakistan	1970	0.00335	0.00590	0.01210	1.75820	3.60902
	1975	0.00212	0.00600	0.01245	2.78650	5.88340
	1980	0.00253	0.00589	0.01298	2.32870	5.12983
	1985	0.00258	0.00589	0.01377	2.16426	5.33310
	1989	0.00212	0.00590	0.01478	2.77731	6.96090
Bangladesh	1970	0.00133	0.00102	0.01284	0.76790	9.62730
	1975	0.00144	0.00102	0.12787	0.71286	8.74005
	1980	0.00136	0.00102	0.01284	0.75230	9.43420
	1985	0.00139	0.00102	0.01333	0.73560	9.56417
	1989	0.00121	0.00100	0.01408	0.82498	11.66769

Sri Lanka	1970	0.00075	0.00045	0.00272	0.66200	3.63890
	1975	0.00062	0.00050	0.00271	0.79855	4.36565
	1980	0.00036	0.00050	0.00280	1.39200	7.83400
	1985	0.00046	0.00050	0.00274	1.07914	5.96490
	1989	0.00034	0.00050	0.00270	1.44830	7.90450
Nepal	1970	0.00027	0.00105	0.00322	3.84340	11.82460
	1975	0.00025	0.00105	0.00317	4.32490	13.09960
	1980	0.00017	0.00105	0.00314	6.24790	18.73540
	1985	0.00019	0.00105	0.00318	5.44170	16.52030
	1989	0.00014	0.00105	0.00334	7.41915	23.66170

1. Y/Yw is the ratio of GDP in country x and the world GDP
2. Vi_{land}/Vw_{land} is the land endowment share of country x i.e. ratio of endowment of land in country x and world endowment of land.
3. Vi_{lab}/Vw_{lab} is the labour endowment share of country x i.e. ratio of endowment of labour in country x and world endowment of labour
4. Land Abundance ratio is: $[(vi_{land}/Vw_{land})/(Yi/Yw)]$
   Country is abundantly supplied in land if this ratio exceeds unity.
5. Labour Abundance ratio is:
   $[(vi_{lab}/Vw_{lab}) / (Yi/Yw)]$

Source :FAO Production Year Book 1986 and 1990 issue. National Accounts Statistics: Main Aggregates: 1990 and 1991.

other countries in the study, GFCF/land ratio was also lowest in South Asia (except Sri Lanka and Bangladesh) as compared to all other countries in study. Similarly skilled labour/unskilled labour ratio was lowest in South Asia (except Sri Lanka) as compared to other countries in the study. However, land to labour ratio was higher in all South Asian countries (except Bangladesh) as compared to Korea and Singapore and not much different from Malaysia, Philippines and Thailand (See Table 4.4). We may therefore, infer that (in general) South Asian countries are relatively abundant in land (except for Bangladesh) and unskilled labour.

4.3.3 Factor Intensity of Agricultural and Labour Intensive Products

The calculation of factor intensity is based on the assumption of strong factor intensity,[23] i.e. that the ranking of industrial factor intensities do not change in the face of a desperate structure of relative factor prices in other countries. The average factor intensity of agricultural products and labour intensive products was established on the basis of Leamer's typology of industries. The typology of industries by Leamer (1984) was based on the average relationship between factor abundance and comparative advantage. According to his simple correlations and an econometric model, all the agricultural and allied products comprising sections 0,1,2 (less 27 and 28) sec. 4, Div. 63, 64 and 94 of SITC [24] are intensive in the use of unskilled labour and land. Light manufactures comprising Div. 66, 82, 83, 84, 85, 89, 91, 93 and 96 of SITC require mainly low skilled labour and are thus classified as labour intensive products and heavy manufactures comprising section 5 and 7 and Div. 61, 62, 65, 67, 69, 81, 86, 87 and 88 are intensive in the use of capital and skill and are thus termed capital intensive products. We assume these industry specific factor intensities to hold in the context of South

Table 4.4: Relative endowment of Capital, land and labour (skilled and unskilled) South Asia, NICs and ASEAN-4.

Country	Year	GFCF/Land	GFCF/Unskilled Lab	Skilled/ Unskilled Lab	Land/ labour
Korea	1970	227.111	162.794	7.065	0.86
	1975	621.957	406.805	7.696	0.75
	1980	2023.831	172.972	11.821	0.67
	1985	2771.056	253.211	14.625	0.59
		(1120.130)	(1455.410)	(107.010)	
Singapore	1970	10823.070	273.734	2.226	0.08
	1975	35754.316	--	--	0.06
	1980	83598.770	189.603	3.444	0.05
	1985	130967.370	438.417	6.194	0.05
		(1110.080)	(1501.620)	(178.260)	
Malaysia	1970	16.913	35.815	1.381	8.89
	1975	71.228	150.221	1.857	7.38
	1980	232.054	469.916	2.290	6.16
	1985	283.467	567.369	2.638	5.32
		(1576.030)	(1484.165)	(91.020)	

Philipines	1970	39.850	49.682	4.747	2.17
	1975	125.547	139.245	4.882	1.89
	1980	302.288	307.923	4.804	1.70
	1985	162.121	170.133		1.50
		(658.565)	(519.795)	(1.201)	
Thailand	1970	32.875	111.980	3.672	2.86
	1975	66.524	203.859	4.405	2.50
	1980	158.140	286.005	7.333	2.17
	1985	172.876	368.770	10.111	1.92
		(425.859)	(229.317)	(175.289)	
Indonesia	1970	6.922	6.244	1.273	3.97
	1975	22.643	21.197	1.596	3.59
	1980	89.900	88.536	2.058	3.22
	1985	1116.545	1386.465	3.348	2.86
		(16030.381)	(22105.825)	(163.047)	

India	1970	28.254	56.834	0.515	1.33
	1975	53.528	10.392	0.590	1.22
	1980	112.394	21.168	0.681	1.12
	1985	147.532	26.479	0.770	1.01
		(422.182)	(365.881)	(49.437)	
Pakistan	1970	19.191	9.567	0.250	3.99
	1975	31.522	14.377	0.299	3.51
	1980	56.308	23.080	0.351	3.03
	1985	71.299	26.198	0.420	2.18
		(271.517)	(173.824)	(68.00)	
Bangladesh	1970	--	--	--	0.65
	1975	75.485	6.058	0.351	0.59
	1980	178.458	13.392	0.409	0.53
	1985	155.665	10.802	0.495	0.46
	1990	--	--	0.545	--
		(271.517)	(173.824)	(68.000)	

Srilanka	1970	61.843	41.867	3.546	1.49
	1975	78.037	--		1.36
	1980	207.799	189.635	6.692	1.19
	1985	214.105	181.500	6.752	1.09
	1990	--	--	7.621	1.02
		(246.208)	(333.519)	(90.425)	
Nepal	1970	--	--	0.149	2.66
	1975	14.769	4.475	0.238	2.45
	1980	22.423	6.293	0.259	2.23
	1985	37.603	10.064	0.344	1.99
	1990	--	--	0.351	1.72
		(154.628)	(124.916)	(44.823)	

GFC : Gross fixed Capital Formation in thousand US $., National Accounts Statistics: Main Aggregates,U.N.

Lab(L) : Total number of Economically Active Population in thousands.

Unskilled Lab(3): Total number of unskilled Lab (in thousands) i.e. L3 = Lx Illiteracy rate

Skilled Lab : Skilled Labour in thousands is L-L3.

Source : Production Year Book, FAO, Vol.40, 1986 and Statistical yearbook, UNESCO, 1980,1985,1990

Note : Figures in brackets show percentage increase in relative factor endowment from 1970 to 1985.

Asia. Broad confirmation of the assumption is available from the results of alternative estimates of factor intensities for India by Sengupta (1990) [25]. The empirical results of the paper showing total labour and capital requirement of various sectors of Indian economy with 14 sector breakdown (presented in Table 4.5) depict that leading labour intensive sectors in India are agriculture and allied sectors. The most capital intensive sector is power followed by other transport and manufacture.

4.3.4 Comparative Advantage: Interaction between Endowment and Intensity

The absolute resource abundance Ratio (Table 4.3) depicted abundance of land and labour as compared to capital in all the four South Asian countries for 1970, 75, 80, 85 and 1989. Further, we may state that based on the assumption of strong factor intensity, agriculture and allied products are intensive in the use of land and unskilled labour and labour intensive manufactures are intensive in the use of low skilled labour. **On the basis of interaction between endowment and intensity (Table 4.6) we may state that comparative advantage for South Asian countries is high in case of agricultural and allied products and labour intensive manufactures but low in case of capital intensive manufactures.**

4.4 Dynamic Nature of Comparative Advantage: South Asia

Balassa's (1965) theory of "The stages approach to comparative advantage" uses the static commodity composition of trade as the reference pattern to gauge anticipated changes in comparative advantage on the basis of changes in country's factor endowment [26]. According to this theory, with development, the

Table 4.5: Estimates of Factor Intensities : India (1979-80)

Sector	Total Labour requirements under existing structure of production per million Rs. of output.	Total Capital requirements under existing structure of production per million Rs. of output.	Total (Direct and Indirect) labour requirements under assumed structure of production when all intermediate inputs are supplied domestically (Man years)	Total (Direct and Indirect capital requirements under assumed structure of production when all intermediate inputs are supplied domestically (Million Rs.)
Agriculture	211.90	7.58	211.0	5.34
Forestry	186.53	2.69	188.2	2.03
Fishing	191.05	10.51	191.2	9.84
Mining	68.77	9.42	68.2	8.03
Manufacture	124.73	18.98	168.0	11.24

Construction	134.39	9.81	127.0	5.00
Electricity	60.7	38.36	60.0	36.71
Railways	123.45	11.27	124.0	9.49
Other transport	144.06	19.35	147.0	12.20
Communication	100.86	7.75	100.3	7.19
Trade and Storage	129.48	13.02	97.5	3.59
Banking	97.33	2.83	67.5	1.93
Real Estate	25.97	7.35	16.3	6.11
Public Administration	162.25	9.36	219.3	11.94

Source : Arun K. Sengupta : Alternative Estimates of Factor Intensity and NetSaving(cost) of Factors: Some Results for Indian Economy, IER, Vol.XXV No.1.

Table 4.6: Interaction Between Factor Intensity and Factor Abundance of Different Product Groups : South Asian Countries.

	Factor Intensity			Factor Abundance			Comparative Advantage
Sector	Capial	Land	Unskilled Labour	Capital	Land	Unskilled Labour	
Agriculture and Allied Products	Low	High	High	Low	High	High	High
Labour Intensive Products	Low	-	High	Low	High	High	High
Capital Intensive Products	High	Low	Low	Low	High	High	Low

relative endowment of reproducible tangible capital expands in relation to other factors. With this change, the comparative advantage shifts from agricultural and raw material exports to labour intensive and then to capital intensive exports. The dynamic comparative advantage (in the present study) is examined by using the same approach. The changes in factor endowment are taken to be reflected in changes in empirically approximated endowment of capital and skilled labour.

Relative endowment of land, capital, skilled labour and unskilled labour (Table 4.3) depicts that between 1970 and 1985, there was significant increase in endowment of capital and skilled labour in case of Korea, Singapore, Malaysia, the Philippines and Thailand. With this increase in factor endowment, comparative advantage of these countries shifted from agricultural and raw material products to labour-intensive and capital intensive products. In the case of South Asia, however, there was insignificant shift in comparative advantage as the increase in endowment of capital and skilled labour, as compared to land and unskilled labour, was quite insignificant during 1970 and 1985. Figures in parentheses of Column 3,4 and 5 of (Table 4.4) depict that the ratio of capital/unskilled labour between 1970 and 1985 increased by around 1500 per cent for Korea, Singapore and Malaysia, above 2000 per cent for Indonesia and 600 percent for the Philippines (1970-80) but only by 365 per cent for India, 273 percent for Pakistan, 106 percent for Bangladesh, 246 percent in Sri Lanka and 154 percent in Nepal. The ratio of GFCF to land increased by more than 1100 percent for Korea, Singapore and Malaysia, about 700 percent for the Philippines (1970-80), 426 percent for Thailand but only by 272 percent in Pakistan, 106 percent in Bangladesh, 246 percent in Sri Lanka, 155 percent in Nepal and 422 percent in India. Similarly ratio of skilled labour to unskilled labour increased by more than 100 percent for Korea, Singapore, Thailand and Indonesia, 91 percent for Malaysia, but by 49 percent in India, 68 percent in Pakistan, 40 percent in Bangladesh, 90 percent in Sri Lanka and 44 percent in Nepal.

Thus the increase in endowment of capital and skilled labour was lowest and quite insignificant for South Asian countries (except Sri Lanka) as compared to other countries in the study. The difference in growth rate of factor endowments would be more significant if we look at the fact that the endowment of capital and skilled labour was lowest for South Asian countries as compared to all other countries in the base year i.e. 1970. The ratio of land to labour, however, was not much different for South Asian countries (except Bangladesh) as compared to all other countries in the study.

4.5 Conclusion

South Asian countries as per the empirical approximation, of the study, are relatively abundant in land and unskilled labour as compared to capital and skilled labour. Thus, the comparative advantage for South Asian countries, as per the factor endowment proposition of comparative advantage (in static as well as in dynamic sense) lies in agricultural and allied products and labour intensive manufactures.

Notes

1. Nayyar, Deepak (1976) : India's Exports and Export Policies in 1960s.
2. James, W.E., Naya, Seiji and Meier, G.M. (1987) : Asian Development.
3. H-O theory of comparative advantage is available in two versions. (a) The Factor Content Version, (b) The Commodity Version. The factor content version depicts trade as an implicit exchange of the factors of production embodied in traded goods. The commodity version suggests that a country will tend to export goods which require comparatively large amounts of those factors of production which are relatively abundant in the country. Thus, a country which is rich in capital would be expected to

produce and export goods which require relatively large amounts of capital.

4. Balassa, Bela (1965): Trade Liberalisation and Revealed Comparative Advantage.

5. In this context the following extracts are of relevance,

"The HOV equations are a set of relations between factor intensities, trade and resource endowments. A proper measure of the accuracy of the theory begins with independent measures of these three concepts and determines the extent to which theory conforms to the HOV equations" (Leamer, 1984).

"The H-O or factor proportions theory of international trade maintains that comparative advantage (and hence the pattern of trade) can be explained on the basis of the products' factor intensities and countries' factor endowments. In other words, the H-O theory of international trade involves three basic elements, the factor intensity of commodities; the factor endowment of countries; and the pattern of trade flows. Based on factor proportions, it shows the pattern of trade in the light of the interactions between factor intensity and factor endowment" (UNIDO, 1985).

"Comparative advantage is the outcome of the interaction between two variables; factor endowment, which is a country characteristic, and factor intensity, which is an industry or product characteristic" (Hirsch, 1974). "Factor intensity and factor endowment they are independent of each other and are exogenous factors. Their interaction produces comparative advantage. Comparative advantage is high when high factor intensity is combined with high factor endowment or when low factor intensity is combined with low endowment. Comparative advantage is low when high factor intensity is combined with low factor endowment or vice versa".

6. See Leamer (1984), UNIDO (1986), Huffbauer (1970) and S.Hirsch (1974). Please see Bibliography for details.

7. Leamer (1984): Sources of Comparative Advantage.

8. Huffbauer, G.C. (1970) : The impact of national characteristics and technology on the commodity composition of trade in manufactured goods.

9. Hirsch, S. (1974) : Capital or Technology? Confronting the Neo-Factor Propositions and Neo-Technology Accounts of International Trade.

10 Leamer (1984): Sources of Comparative Advantage.

11. UNIDO (1986) International Comparative Advantage in Manufacturing.

12 Leaner, (1984), op.cit.

13. Hirsch (1974); Capital or Technology ? Confronting the Neo-factor Proportions and Neo-Technology Accounts of International Trade.

14. Leamer (1984), op.cit.

15. Panchmukhi, V.R. (1978) : Trade Policies of India.

16. Hirsch, S. (1974) : Capital or Technology.

17. Leamer (1984), op.cit.

18. Panchmukhi (1978): Trade Policies of India.

19. Ballance, Robert, H. (1987): International Industry and Business.

20. The concept of "Factor Orientation of RCA" was introduced by a U.N. study to describe and analyse the relationship across countries between factor abundance and Comparative Advantage (CA). Factor orientation may be understood to be a descriptive statistic in the form of a correlation between RCA and factor endowment. By measuring the concordance between the empirical distribution of resources and the pattern of trade performance among countries, the indices give some indication of the relationship between factor abundance and comparative advantage for each of the product groups under study. The values of factor orientation indices range between-1 and +1. Through these orientation indices a typology of industries is established, based on the average relationship between factor abundance and CA. For example, one industry is identified for which relatively high endowments of physical capital appear to be a precondition for a country to develop CA. Another shows the same characteristics in respect to human capital and technological capability. The H-O hypothesis that, in general, CA results from the interaction between high relative factor intensity and abundance of the same factor is tested empirically by relating the factor orientation indices to actual factor intensities. The comparison between the factor orientation of trade in manufactures and the actual factor endowment of a given country allows conclusions to be drawn regarding the adequacy of commodity composition of a country's exports and imports in relation to its factor endowment.

21. (a) Huffbauer, G.C. (1970) : The Impact of National Characteristics and Technology on Composition of Trade in Manufactured Goods.

(b) Balassa, Bela (1989) : Comparative Advantage, Trade Policy and Economic Development.

22. UNIDO (1985); Industry in the 1980s.

23. The validity of the assumption that the capital intensity rankings are irreversible in different countries has been questioned by several authors without any clear subsequent consensus. One of the theoretical possibility that was suggested for explaining Leantif's paradox was comprised by the factor intensity reversals by Minhas.

Minhas's report (1962) stated that estimation of homohypallagic production functions reveals high possibility of factor intensity reversals. Most of the subsequent studies of Minhas's however tend to point to the refutal rather than the confirmation of factor intensity reversals. An empirical test of whether the assumption of factor intensity reversal is true in practice was done by C.Kim (1983). The results of his empirical test (for Germany) not only rejected the hypothesis of factor intensity reversals, but affirmatively supported the hypothesis of strong factor intensity.

24. SITC-Standard International Trade Classification.

25. Sengupta (1990): Alternative Estimates of Factor Intensities for India, Indian Economic Review.

26. UNIDO (1986) ; International Comparative Advantage in Manufacturing.

Chapter 5

Comparative Advantage and Pattern of Trade

5.1 Introduction

The empirical verification of South Asian countries' pattern of trade (Chapter 4) revealed that as per the factor endowment proposition of comparative advantage (in static as well as in dynamic sense), South Asian countries have comparative advantage in agricultural products and labour intensive products. An attempt is made in this chapter to examine the extent to which South Asian countries' pattern of trade conforms to this structure of comparative advantage i.e. to examine the proposition as to whether the agricultural exports are in conformity with present and potential comparative advantage of South Asian countries. The analysis proceeds as follows:

First: Revealed Comparative Advantage (RCA) an index of comparative advantage - is worked out for South Asian countries, NIC-4 and ASEAN-4 for measuring comparative advantage in these countries.

Second: Comparison between ideal comparative advantage (based on resource endowment structure) and the RCA (based on observed pattern of trade) is made to see the concordance between the two. The concordance between ideal comparative advantage and RCA is necessary if we want to use RCA index as an index of comparative advantage for intercountry and intersectoral comparisons.

Third: Composition of exports are worked out for South Asia, NIC-4 and ASEAN-4.

Fourth: Intercountry and intersectoral comparisons of composition of exports vis-à-vis comparative advantage are made to observe whether the composition of trade has been as per comparative advantage.

Before we proceed to empirical analysis, the concept of RCA and commodity composition of exports is given in the following paragraphs.

5.2 Revealed Comparative Advantage: A Measure of Comparative Advantage

5.2.1. Basic Concepts

Empirical verification of South Asian countries pattern of trade as per resource endowment structure revealed that ideal comparative advantage for these countries lies in agricultural products and labour intensive manufactures. The rigorous analysis of ideal comparative advantage, however, requires cost comparisons across countries. The data required for such analysis are rarely available as the concept of comparative advantage is usually defined in terms of autarkic price relationships and all empirical economic data are based on events in the world of trade. As a consequence, "true" indices based on pretrade prices cannot be obtained and all "measures" of comparative advantage have to be considered as approximations of the underlying "true" relationships.

Three alternative approaches have been suggested for estimating comparative advantage: [1]

(a) Input-output technique;

(b) Factor Intensity measured by elasticities of substitution; and

(c) Export performance index.

With regard to first and second approach, a number of criticisms have been offered. The most common among all is that these have over emphasised cost elements ignoring thereby the consideration of non-price factors in the determination of comparative cost advantage. As pointed out by Balassa, cost considerations will not be sufficient... and a complete explanation of comparative advantage would have to take account of the effects of non-price factors".

The export performance index is an indicator which reflects all types of costs and non-price factors "like goodwill, quality, service facilities etc. and is based on empirical data pertaining to trade flows" [2]. Since this pattern in comparative advantage is revealed by the observed pattern of trade flows, it is called revealed comparative advantage [3]. Our study will consider this consider this ratio for assessing comparative advantage.

So far, two types of trade indices have been developed by economists.

5.2.1.1 Trade-cum-Production Index: RACGO and BOWEN Indices: Trade-cum-production indices comprise the four obvious and directly observable quantitative relations used i.e. ratios of:

(a) Exports to Production (X/P)

(b) Imports to Consumption (M/C)

(c) Net Trade to production (X-M/P)

(d) Production to Consumption (P/C)

A major problem with trade-cum-production indices is that for many products and countries production data are not available. Even when they are available they often cannot be concorded with data on imports and exports. The alternative left is to use trade only indices.

5.2.1.2 Trade only Indices : Trade only indices are solely based on trade data and comprise:

(a) RNX and Donges and Riedel Index: The ratio of net exports to the sum of exports and imports (T/XM).

(b) Bela Balassa Index: Ratio of the share of a given industry in total manufacturing exports of the country under consideration to the corresponding share for the trading world as a whole.

In view of the impossibility of specifying which index is best, and also of the alternative interpretations that may be given to them, the question of consistency among the indices is of considerable relevance. Three types of consistency tests were employed by U.N. [4] to accord with the three different ways of interpreting the revealed comparative advantage indices. These were, simple correlations or cardinal measures: Rank correlations or ordinal measures and counts of consistent versus inconsistent indications of comparative advantage (or comparative disadvantage) by dichotomous measures.

The results of the tests indicated the following:

The trade-cum-production indices proved to be acceptably consistent in the cardinal as well as in the ordinal sense. Moreover, they were perfectly consistent as dichotomous indicators of comparative advantage. The trade only indices were not very consistent as cardinal measures. However, a higher degree of consistency was achieved when used as ordinal interpretations and very high when they were used as dichotomous interpretations.

Thus far, at least three different ways of interpreting an index of comparative advantage have been mentioned. The first one is the "dichotomous" interpretation whose objective is to determine whether comparative advantage (vis-a-vis the "rest of the world") prevails for a given product and a given country. A second one is the "ordinal" interpretation, which employs an index to rank countries by their degree of comparative advantage in a particular product. The third and most stringent one is the "cardinal" interpretation, which allows conclusions to be drawn concerning differences in comparative advantage in terms of a notion of "size'.

Any approach to an empirical assessment of patterns of comparative advantage is largely determined by the purpose of the assessment and the suitability of the data available. In the present case, our objective was to draw a comprehensive version of the South Asian countries competitiveness for different product groups as compared to NICs and ASEAN-4. The use of trade data as a basic source of information was therefore almost predetermined.

Such data are not only comparable for countries and industries (or product groups) and time frames, they are also acceptably detailed with respect to product categories. While the latter characteristic reduces the risk of patterns of comparative advantage becoming obscured through aggregation, the former allows for wide ranging comparisons to be made in terms of countries and time. These comparisons assume added importance when attempts are being made to identify generally valid rules concerning the determination of comparative advantage by country and industry characteristics. Thus, it becomes pertinent to use trade only indices for the inter-sectoral and inter-country comparisons of comparative advantage vis-a-vis composition of trade of South Asian countries as compared to NICs and ASEAN-4.

The most popular trade only index is Bela Balassa's Export Performance Ratio. According to Balassa (1965) a country's comparative advantage (CA) in the trade of a particular commodity can be measured by the share of that commodity in the country's total exports relative to the commodity's share in total world exports, i.e.

$RCA_{ij} = (X_{ij}/X_j) / (X_{iw}/X_w)$ I

where

RCA_{ij}-denotes revealed comparative advantage (RCA) of country j for product i;

X_{ij} -the value of country j's exports of commodity i

X_j -total commodity exports of country j

X_{iw} - value of world exports of commodity i

X_w - value of total world exports.

If ratio RCA_{ij} is greater than unity, it is generally interpreted to mean that the country is at a comparative advantage in the export of the product in question. However, if the RCA index is less than unity it is taken to indicate that the country is at a comparative disadvantage in the export of that product.

5.2.2 The Commodity Composition of Exports

The commodity composition of exports has been structured under four groups namely Raw Materials (fuel and non fuel), agricultural products, Labour Intensive manufactures and Capital intensive manufactures. The grouping used in the text is based on ten commodity aggregates by Leamer (1985) [5]. The ten commodity aggregates are petroleum (PETRO), Raw Materials (MAT), Forest Products (FOR), Tropical Agriculture (TROP), Animal Products (ANL), Cereals (CER), Labour Intensive Manufactures (LAB), Capital Intensive Manufactures (CAP), Machinery (MACH) and Chemical (CHEM). Leamer grouped the ten commodity aggregates into three groups. The first group comprised of the nature resource product: PETRO and MAT, which are expected to be related to the nature resources coal, minerals and oils. The next four aggregates: FOR, TROP, ANL and CER are grouped as crops and are expected to be positively related to the land variables. The last four aggregates: LAB, CAP, MACH and CHEM are grouped as manufactured commodities and are expected to be related positively to capital and labour. We have subgrouped Leamer's third group i.e. manufactures into two i.e. (1) Labour Intensive Manufactures and (2) Capital Intensive Manufactures. Thus the four groups in our study are :
(1) Raw Materials, (2) Agricultural Products - comprising FOR, TROP, ANL and CER, (3) Lab Intensive Manufactures (LAB) and (4) Capital Intensive Manufactures - comprising CAP, MACH and CHEM [6]. The composition of these aggregates, as per two digit SITC aggregations is described in Table 5.1.

Table 5.1 : Composition of Trade Aggregates

Group S.No.	SITC Division	Leamer's Aggregates	Aggregates in the present study
I	33	Petroleum (PETRO)	Raw Materials (Fuel+non-fuel)
II	27,28,32,34,35,68	Raw Material ((MAT)	
III	24,25,63,64	FOREST Product (FOR)	Agricultural products
IV	05,06,07,11,23	Tropical Agriculture (TROP)	
V	00,01,02,03,21,29, 43,94	Animal Products (ANL)	
VI	04,08,09,12,22,26, 41,42	Cereals etc. (CER)	

VII	66,82,83,84,85,89,91, 93,96	Labour-Intensive Manufactures (LAB)	Labour Intensive Manufactures
VIII	61,62,65,67,69,81	Capital Intensive Manufactures (CAP)	Capital intensive Manufactures (Cap. Intensive +Cap. and Skills intensive)
IX	71,72,73,86,95	Machinery (MACH)	
X	51,52,53,54,55,56,57, 58,59	Chemicals	

5.3 Empirical Results

5.3.1 Revealed Comparative Advantage (RCA)

The indices of RCA - Balassa's export performance ratios - were computed for agricultural products, labour-intensive products, and capital intensive products for South Asia, NIC-4 and ASEAN-4 (Table 5.2).

The RCA indices (Table 5.2) indicate that till the beginning of 1970s, South Asia, NIC-4 and ASEAN-4 all had comparative advantage in agricultural products. The RCA index for these products was above unity for all these countries. A fall in this index for agricultural products and an increase in this index for labour-intensive products was observed from mid-1970s for Korea and Singapore and beginning of 1980s for Malaysia, Philippines and Thailand. Further, there was slight increase in the index for capital intensive products for Korea since mid 1970s and for Singapore, Malaysia, Philippines and Thailand from beginning of 1980s. However, in the case of Korea increase in index of capital intensive products was at the cost of fall in index of labour intensive products (along with primary products). In the case of ASEAN-4 the increase in index of capital intensive products was accompanied by fall in RCA index of primary products only.

In the case of South Asian countries, the change in RCA index was insignificant. There was marginal increase, instead of decrease, in this index for agricultural products from 1.35 in 1970 to 1.96 in 1985 (India), 2.01 in 1970 to 2.26 in 1985 (Pakistan). For capital intensive products, there was marginal decrease from 0.91 in 1970 to 0.85 in 1985 (India) and 0.96 in 1970 to 0.95 in 1985 (Pakistan). However, there was some increase in this index for labour intensive products from 0.26 in 1970 to 2.18 in 1985 (India) and 1.26 in 1970 to 1.43 in 1985 (Pakistan). In Sri Lanka, there was marginal decline in this index from 3.7 in 1970 to 3.58 in 1985 for agricultural products, and some increase in the index for labour intensive products from 1.30 in 1970 to 2.90 in 1985. The index for capital

Table 5.2 : Revealed Comparative Advantage

Trade Group	Year	Korea	Singapore	Malaysia	Philippines	Thailand	Indonesia	India	Pakistan	Bangladesh	Sri Lanka
Agriculture and Allied Products	1966	1.566	1.739	2.403	2.263	3.260	--	--	--	--	--
	1970	1.051	1.773	2.336	2.823	2.863	2.085	1.35 (1)	--	--	3.706 (1)
	1975	0.944	1.087	3.437	3.173	3.520	0.931	1.62 (1)	2.011 (1)	1.594 (1)	3.887 (1)
	1980	0.694	1.145	2.824	2.684	3.494	1.313	1.99 (2)	3.028 (1)	2.304 (1)	3.916 (1)
	1985	0.329	0.847	2.600	1.738	1.458	2.143	1.96	2.263 (1)	2.065 (1)	3.587 (1)
Labour Intensive Products	1970	5.697	1.071	0.384	0.097	0.990	1.476	0.26 (3)	--	--	--
	1975	5.761	1.000	0.381	0.902	0.969	--	1.67 (2)	1.269	--	1.308

	1980	3.328	1.342	0.284	2.952	1.436	0.049	2.18 (1)	0.748	--	1.551
	1985	2.858	1.357	0.326	5.724	1.715	0.220	--	1.430	1.964	2.906
Capital Intensive Products	1970	0.471	0.720	0.117	0.043	0.057	0.018	0.91 (2)	--	--	0.020
	1975	0.828	0.741	0.258	0.053	0.150	0.017	0.87 (3)	0.964	1.431	0.020
	1980	1.129	0.810	0.326	0.151	0.326	0.026	0.85 (3)	0.943	1.498	0.004
	1985	1.223	0.850	0.428	0.236	0.541	0.58	--	0.951	0.937	0.0101

intensive products, in case of Sri Lanka, has however, been insignificant for throughout the period i.e. 1970 to 1985.

The changes in RCA index were consistent with the hypothesis that with economic development, comparative advantage shifts from agricultural products to labour intensive manufactures, to skill and capital intensive manufactures. The shift in comparative advantage in Korea, Singapore and ASEAN-4 was as per increases in relative endowment of capital and skilled labour in these countries. In case of South Asia, specially India and Pakistan, there was insignificant shift in comparative advantage as the increase in endowment of capital and skilled labour was lowest and quite insignificant as compared to other countries in the study for the period 1970-1985 (See Chapter 4). [Section 4.4 and Table 4.5]

5.3.2 Ideal Comparative Advantage and Revealed Comparative Advantage

The empirical verification of H-O model of trade for South Asian countries (Chapter 4) revealed that the ideal comparative advantage (based on interaction between factor endowment and factor intensity) for these countries lies in agricultural products and not in capital intensive products. Hence the composition of exports of these countries, as per H-O proposition, should reflect higher percentage of agricultural exports as compared to capital intensive manufactures. The rigorous analysis of patterns of ideal comparative advantage, however, requires cost comparisons across countries and the data required for such an analysis are rarely available. This is the concept of RCA (based on observed pattern of trade) is used for most of the economic analysis. Moreover, comparison between ideal comparative advantage and RCA reveal (in the present study) concordance between the two in case of South Asian countries. The RCA index, in accordance with the results of interaction between factor abundance and factor intensity (Table 4.3), was above unity for agricultural and allied products and below unity for capital intensive products depicting comparative advantage for these countries in agricultural and allied products in 1970s and 1980s. The consistency between ideal comparative advantage and RCA, may be due to the fact that the difference in the two concepts results from differences in trade policies for agriculture. The revealed comparative advantage is the net result of (a) the ideal comparative advantage determined by the resource endowment structure and (b) trade policies. In the case of NIC-4 and ASEAN-4, the trade policies have been as per the resource endowment structure and in South Asian countries at least till end of 1980s, the policy thrust has been to encourage the export of non-traditional commodities by various policy measures i.e. subsidies, import entitlement and replenishment schemes etc. The traditional sector has not received much policy stimuli which implied insignificant difference between RCA index and ideal comparative advantage. Thus, in order to evaluate the extent to which South Asian countries pattern of trade conforms to the structure of comparative advantage i.e. to examine as to whether the agricultural exports are in conformity with present and potential comparative advantage of South Asian countries comparative advantage, inter-country and intersectoral comparisons of composition of exports vis-à-vis RCA are worked out in the following analysis.

5.3.3 Composition of Exports: NIC-4, ASEAN-4 and South Asia

The NIC-4 opted for export orientation by beginning of 1960s. The composition of exports of NIC-4 (Table 5.3) indicates that exports of agricultural commodities constituted more than 45 per cent of total exports in Korea and 50 per cent of total exports in Singapore. By mid 1960s, their share started declining but still accounted for more than 40 per cent for Korea and 45 per cent for Singapore. By mid 1970s, their share decreased to 20 per cent for Korea and around 25 per cent for Singapore. The share of labour intensive expects was around 4 per cent in the beginning of 1960

Table 5.3 : Composition of Exports: South Asia, NICs and ASEAN-4

Aggregate	Year	Korea	Singapore	Malysia	Philippines	Thailand	Indonesia	India	Pakistan	Bangladesh	Sri Lanka	Nepal
Petroleum and Raw Materials. SITC Div.: 27,32,33, 34,35, 68	1962	19.82	17.68	--	--	--	--	7.74	--	--	--	--
	1966	11.90	24.58	30.60	2.52	12.19	--	11.36	--	--	--	--
	1970	6.68	24.65	29.34	22.51	14.86	44.24	12.67	--	--	0.663	--
	1975	--	--	--	--	--	--	--	1.40	Nil	5.325	--
	1980	1.06	31.12	34.78	21.23	13.94	75.75	7.84	7.40	Nil	18.693	--
	1985	3.56	29.10	33.68	11.40	4.96	78.58	29.83	14.20	2.5459	11.755	0.1758
Agricultural and allied Products SITC Div: 00,01,02,03,04,05, 06,07,08,09,11,12, 21,22,23,24,25,26, 29,41	1962	45.27	49.90	--	--	--	--	47.58	--	--	--	--
	1966	40.16	44.58	64.62	83.68	83.59	--	40.15	--	--	--	--
	1970	27.26	40.02	60.63	73.25	74.28	54.09	34.91	--	--	96.173	--
	1980	20.01	23.04	72.43	67.24	74.60	19.73	34.27	42.921	33.790	82.395	--
	1985	11.45	18.91	46.66	44.31	57.52	21.67	32.91	49.988	38.05	64.660	44.32
		5.02	12.92	39.65	33.03	55.05	22.23	28.09	34.521	31.492	55.693	24.79
Labour intensive Products SITC Div: 66,82,83,84,84,85, 89,91,93,96	1962	3.59	3.97	--	--	--	--	1.78	--	--	--	--
	1966	23.94	6.85	1.39	0.39	1.20	--	5.16	--	--	--	--
	1970	43.34	7.96	2.85	0.72	7.36	10.98	1.95	--	--	0.392	--
	1975	37.79	6.56	2.50	5.92	6.35	--	10.97	8.322	0.559	8.582	--
	1980	30.19	12.18	2.25	26.78	13.02	0.45	19.76	6.788	1.574	14.070	6.359
	1985	25.92	12.31	2.96	51.91	15.55	2.00	--	12.971	17.807	26.350	21.545
Capital intensive Products SITC Div: 61,62,65,69,81,95, Total of sections 5 and 7	1962	10.94	20.51	--	--	--	--	40.40	--	--	--	--
	1966	23.88	21.71	4.66	1.16	1.55	--	43.16	--	--	--	--
	1970	22.82	20.34	5.70	2.11	2.74	0.85	44.03	--	--	0.977	--
	1975	38.39	34.35	11.96	2.47	6.99	0.79	40.21	44.709	66.338	9.405	--
	1980	54.37	35.82	14.39	6.67	14.43	1.17	37.65	41.695	66.203	1.769	23.230
	1985	61.96	43.04	21.70	11.92	27.42	2.95	--	48.195	47.440	5.139	11.791

(except Hong Kong) increased to more than 30 per cent in the 1970s and then fell to less than 30 per cent by the middle of 1980s. As for chemicals, machinery and capital intensive products, the share was around 20 per cent in the 1960s, around 30 per cent in the 1970s and more than 40 per cent in the later 1970s and 1980s.

ASEAN-4 shifted to export orientation by middle of 1960s. Being rich in agricultural and natural resources, these economies specialised in agricultural commodities and raw materials. Tropical products, cereals and raw materials were the three major exports aggregates of ASEAN-4. Agricultural commodities comprised 60 to 80 per cent of total exports from mid 1960s to mid 1970s (Table 5.3). A decline started from late 1970s and this have varied between 45 to 58 per cent in 1980. The share of labour intensive exports in total exports was around 1 per cent in mid 1960s, less than 2 per cent for Philippines and 12 per cent for Thailand in late 1970s and still more to more than 50 per cent for Philippines and 16 per cent for Thailand by middle of 1980s. The share of chemicals, machinery and capital intensive exports was lesser than 1 per cent till middle of 1960s, increased to around 4 per cent by beginning of 1970s (1972) and around 2 per cent by middle of 1980s (1985).

In South Asia (unlike the case of NIC-4 and ASEAN-4), the capital intensive exports were more important than agricultural exports in India, Pakistan and Bangladesh since middle of 1960s. Sri Lanka and Nepal, however, specialised in agricultural and raw material products. Table 5.3 depicts that the share of capital intensive products in total exports of India, Pakistan and Bangladesh were more than 40 per cent, till the middle of 1970s. By beginning of 1980s, though there was slight increase in the relative share of labour intensive exports in India and Pakistan the share of agricultural exports showed little decline. The share of agricultural exports was around 30 per cent in 1970s and less than 30 per cent in 1980s in India. In Pakistan though their share was above 40 per cent in 1970s, it was mainly due to cereal exports, other agricultural product exports were quite insignificant. By middle of 1980s, however, the share of cereal exports too came down, leading to fall in the share of total agricultural exports in Pakistan to 34 per cent only. The share of capital intensive exports and labour intensive exports increased at the cost of cereals in 1985.

5.4 Composition of Exports vis-à-vis Comparative Advantage

5.4.1 Intersectoral Comparisons

Intersectoral comparisons of South Asian countries composition of exports vis-à-vis RC depict that the share of capital intensive exports in India, Pakistan and Bangladesh were always higher than the share of agricultural exports and were comprising around 40 per cent or more of total exports of India, Pakistan and Bangladesh till the beginning of 1980s. Ranking of the three broad product groups [7] as per their RCA and composition of exports for India, Pakistan and Bangladesh reveals that the capital intensive products get the third rank as per RCA but first rank as per the composition of exports for the period 1970 to 1985. On the other hand, agricultural products get the first rank as per RCA but second as per composition of exports for the period 1970 to 1985. Though the share of labour intensive exports showed a little upward trend during 1980s, the share of agricultural exports in these countries showed decline.

Sri Lanka is the only South Asian country which specialised in agricultural and raw material products. The share of agricultural products in total exports of Sri Lanka accounted for 96 per cent in 1970, 82 per cent in 1975, 65 per cent in 1980 and 56 per cent in 1985 (Table 5.2). There was some decline in share of agricultural exports and a certain amount of manufacturing exports has taken place in 1970s and 1980s. Agricultural export sector, however, is still the leading sector in the economy. Plantations are still the major export earners. When the world price remains high, tea, rubber, coconut account for more than 70 per cent of the island's export earnings. Moreover, the increase in the amount of manufacturing exports is also achieved by adding value to their traditional products, such as bagging and packing. In account they have started adding value in shape of fibre. They are also trying for a higher clove price in the world markets.

5.4.2 Intercountry Comparisons

Comparing export performance ratios (RCA) and composition of exports of South Asian countries with those of NIC-4 and ASEAN-4 (Tables 5.1 and 5.2) indicate the consistency between the composition of exports and RCA in the case of NIC-4 and ASEAN-4 and inconsistency for the same in the case of South Asian countries (except Sri Lanka).

The RCA indices (Table 5.2) indicate that till the beginning of 1970s, South Asia, NIC-4 and ASEAN-4, all had comparative advantage in agricultural products. With development, the relative endowment of capital and skilled labour increased in Korea, Singapore and ASEAN-4 and thus the comparative advantage (RCA) for Korea, Singapore and ASEAN-4 countries shifted, in the first phase, from agricultural products to labour intensive products (beginning of 1970s for Korea, late 1970s for Korea, late 1970s for Singapore, Malaysia, Philippines and Thailand) and to capital intensive products in the second phase (mid 1970s for Korea and beginning of 1980s for Malaysia, Philippines, Thailand and Singapore). In accordance with this shift in comparative advantage the composition of exports shifted from primary exports to labour intensive and then to capital intensive exports. The composition of exports (Table 5.3) indicates that exports of agricultural commodities increased in the beginning of 1960s and constituted more than 45 per cent of total exports in Korea and 50 per cent of total exports in Singapore. By mid 1960s, their share started declining but still accounted for more than 40 per cent for Korea and 45 per cent for Singapore. By mid 1970s, their share decreased to 20 per cent for Korea and around 25 per cent for Singapore. Similarly in ASEAN-4 countries (except Indonesia) agricultural commodities comprised 60 to 80 per cent of total exports from mid 1960s to mid 1970s. A decline started from late 1970s and this share varied between 45 to 58 per cent in 1980.

In the case of South Asian countries, the relative endowment of capital and skilled labour was lowest and quite insignificant as compared to other countries in the study and thus the change in comparative advantage (RCA index) was insignificant. There was marginal increase, instead of decline, in this index for agricultural products from 1.35 in 1970 to 1.96 in 1985 (India), 2.01 in 1975 to 2.26 in 1985 (Pakistan), 1.59 in 1975 to 2.06 in 1985 (Bangladesh) and marginal decline from 3.70 in 1970 to 3.59 in 1985 (Sri Lanka). For capital intensive products there was marginal decrease from 0.91 in 1970 to 0.85 (India), 0.96 in 1975 to 0.96 in 1985 (Pakistan), 1.43 in 1975 to 0.93 in 1985 (Bangladesh) and 0.02 in 1970 to 0.01 in 1985 (Bangladesh) and 0.02 in 1970 to 0.01 in 1985 (Sri Lanka). The composition of exports of South Asian countries (except Sri Lanka) unlike NIC-4 and ASEAN-4, were not consistent with this insignificant change in RCA index from 1970-1985. The share of agricultural exports declined from 35 per cent in 1970 to 28 per cent in 1985 and further to 17 per cent by 1990 (India), 43 per cent in 1975 to 35 per cent in 1985 (Pakistan) and 34 per cent in 1975 to 31 per cent in 1985 (Bangladesh). Per capita exports of agricultural products were very low in South Asia (except Sri Lanka) compared to NICs and ASEAN-4 countries.

5.5 Conclusion

The overall industrialisation and trade strategy of the NIC-4 and ASEAN-4 supported an output mix that conformed to domestic factor endowments and hence comparative advantage. Korea, Singapore and ASEAN-4 countries, having realised that an expanding industrial sector cannot be fuelled for long by resource transfers from a static agricultural sector, improved their competitiveness in world trade by mobilising potential of agriculture in the initial stages of development. These countries, following the Bela Balassa's theory of shifting comparative advantage, relied on high percentage of primary exports (mainly agricultural) in the initial stages of development (Korea in early 1960s, Singapore and ASEAN-4 in late 1960s and early 1970s), shifting to labour intensive exports (mainly resource based) in the second stage (Korea in the early 1970s and ASEAN-4 in late 1970s) and highly skilled and capital intensive exports in the third stage (Korea in the early 1980s and ASEAN-4 and Singapore by mid 1980s).

In South Asian countries (except Sri Lanka), the industrialisation and trade strategy, in a haste to catch up with the world, aimed at industrial self-sufficiency (especially in basic industries, e.g. chemicals, metal products and machinery) while disregarding the country's factor endowments and bottlenecks to industrial development. the comparative advantage for South Asian countries, as per neo-classical doctrine, lies in agricultural products and labour products. The composition of exports of these countries (especially India, Pakistan and Bangladesh), however, depict higher percentage of capital intensive exports in the initial stages of development. Though, as regard labour intensive exports there was some awareness on the part of these countries and the share of labour intensive exports showed little upward trend from beginning of 1980s, the share of agricultural exports showed decline not only in 1970s but even in 1980s. There is considerable potential for increasing agricultural exports from South Asia. In a nutshell, we may state an increase in the share of agricultural exports in the export basket of these countries may, thus, lead to better performance of their exports. In other words, performance of South Asian countries exports may be improved by realising greater foreign exchange earnings and improving the linkage between agriculture and other sectors by expanding agricultural exports.

Notes

1.Panchmukhi (1978).

2.Balassa, Bela (1967).

3.Panchmukhi (1978).

4. (a) The objective of the *dichotomous interpretation* is to determine whether comparative advantage or disadvantage (vis-à-vis the rest of the world) prevails for a given product and a given country.

(b) *Ordinal Interpretation* employs an index to rank countries by their degree of comparative advantage in a particular product;

(c) *The Cardinal Interpretation* allows conclusions to be drawn concerning differences in comparative advantage in terms of a notion of size.

5. The ten commodity aggregates are formed from 61 two digit SITC categories on the basis of two method (1) an algorithm based on the correlation matrix alone. The aggregation analysis begins with the 61x61 matrix of cross country correlations of the net export data using trade data from as many as 60 countries. A high correlation in a given year between two commodity classes indicates that these commodities behave similarly in international trade in the sense that if a country has large positive net exports of one, then it also has large net exports of the other. When such a high correlation is found, these classes are combined into one, since the forces that determine trade in a component are likely to be the same as those that determine trade in the corresponding aggregate. (2) A set of cross section regressions of the export data on a list of resources, and commodities are aggregated that have similar regression coefficients.

6. Machinery and chemicals are not only capital intensive but skill intensive too which requires still more capital in the form of human capital.

7. We show ranks of three broad product groups i.e., agricultural, labour intensive and capital intensive products in South Asian countries in brackets of Table 4 and 5.

Chapter 6

Strategy for Agricultural Exports: India

6.1 Introduction

Empirical verification of South Asian countries' pattern of trade (Chapter 4) revealed that as per the factor endowment proposition of comparative advantage, South Asian countries have comparative advantage in agriculture and allied products and labour intensive manufactures. The analysis of inter-country and inter-sectoral comparisons of composition of exports vis-à-vis comparative advantage of South Asian countries (Chapter 5), however, revealed that agricultural exports of these countries do not conform to the structure of their comparative advantage as reflected in the resource endowment structure. It was, thus concluded that an increase in the share of agricultural exports and labour intensive exports in the export basket of these countries, may lead to a better performance of their exports. The question arises : How to increase agricultural exports of these countries? In other words, what should be the strategy of agricultural exports for these countries? Before we proceed to analyse the strategy of agricultural exports, it becomes imperative to analyse the importance of an agricultural export strategy for developing South Asian countries and the present position, performance and problems of agricultural exports of these countries. The importance of the agricultural export strategy reflects the need for agricultural export strategy and the performance of agricultural exports of these countries would reflect the feasibility of increasing agricultural exports. The feasibility of increase in agricultural exports may be worked out by observing the performance of agricultural exports of these countries (a) as per their sectoral output, and (b) as per the world scenario. Thus an attempt is made in the present chapter, by taking a case study of India, [1] to suggest a suitable strategy of agricultural exports for these countries. The analysis proceeds as follows:

(a) Importance of agricultural export strategy for developing countries.

(b) Importance of agricultural exports : India;

(c) Present position, performance and problems of India's agricultural exports; and

(d) Strategy of India's agricultural exports.

6.2 Importance of Agricultural Export Strategy for Developing Countries

Developing countries are currently at a cross road. They are simultaneously facing major short term adjustment problems of critical foreign exchange shortages and the prospects of a continuation of a less favourable international environment in which to carry out these adjustments.

The present plight of the developing countries according to Irma Adelman (1984) [2] is the result of adverse changes in the external environment since 1973 and of the policy choices made by developing countries themselves. The adverse conditions in the international environment were: (a) The slowdown in the volume of world trade due to the two global recessions in the OECD countries during the 1970s; (b) The two oil price shocks of 1980s; and (c) The protectionist sentiment and protectionist legislation introduced in OECD countries. Despite these adverse conditions in the international environment, developing countries chose to continue placing primary emphasis on rapid industrialisation which required increase in imports of capital goods, technology and food from the OECD countries. The financing of these imports, not only from exports, but also by increasing debts and decumulating reserves resulted in an increase in the debt service obligations of less developed countries. The debt service obligations increased to such an extent that the access of LDCs to commercial credit to support exports was severely decreased.

These developments, combined with the need to revise downward projections for the growth in world demand for non-fuel merchandise exports from LDCs, in the views of Irma Adelman [3], call for a reassessment of the feasibility of continuing to rely on manufacturing export led growth as the major development dynamic for most LDCs during the next decade. Most LDCs which do not have established export markets are unlikely to be able to break into international markets for non-traditional exports in the next decade [4].

Thus, the need to search for an alternative strategy is acute for LDCs. Two strategies - Agricultural demand led industrialisation by Irma Adelman [5] and Balanced growth by Hans Singer [6] - may be of relevance in this context.

According to Irma Adleman (1984) : "In searching for alternative strategies, one should distinguish between an open-development strategy, in which trade is a handmaiden of growth and trade incentives are neutral and an export led strategy in which trade is the source of growth and trade incentives are biased in favour of exports. Export led growth is not the only potentially promising open development strategy. A reallocation of investment resources within the economies of LDCs in favour of an open development strategy, which is agriculturally driven rather than export driven, can lead to superior results under the economic conditions of the coming decade. It should be emphasised that this strategy, named as an agricultural demand led industrialisation (ADLI) strategy is not an argument for a closed development strategy. It is rather a call for a shift in sectoral emphasis for public investment while continuing to maintain or even, if necessary, to switch to an open development strategy".

In analysing the prospects for attaining the Lima target, which stipulated that industrial production in developing countries should rise from 7 percent of world industrial production to 25 percent by the year 2000, Hans Singer advocated a similar strategy as that of ADLI which he called a balanced growth approach. He described this strategy as one of basing "national development on agriculture as the primary sector and developing industries with strong emphasis on agriculture industry linkages and interactions" [7]. This strategy would stress the raising of agricultural productivity, especially that of medium scale farmers as a means of achieving industrialisation. It would accomplish the industrialisation goal by expanding internal demand for intermediate and consumer goods produced by domestic industry. In arguing for this approach, Singer emphasised the linkage effects of agriculture [8]. Along with the emphasis of linkage effects of agriculture, Adelman goes further to emphasise the change in relative importance of these effects with economic development. According to him, "though agriculture can have important linkage effects with manufacturing and the role of agriculture in economic development in functioning as a provision of a surplus for investment in manufacturing, a source of labour, a source of food, and a source of demand linkages, has been recognised by economists, it has not been recognised that the relative importance of these functions changes dynamically with the economy's development. According to him, after the very initial stages of industrial development, the emphasis in policy toward agriculture should shift from surplus extraction to surplus creation and to generation of demand linkages with the rest of the economy". (Adelman, 1984) [9].

The ADLI strategy advocated by Irma Adelman and balanced growth approach by Hans Singer is strengthened by the empirical studies of Banerjee (1975) and James, Naya and Meier (1987). According to Banerjee (1975) [10],

"It is not implied that industrialisation and the expansion of industrial exports should be at the expense of the agricultural development of a LDC. On the contrary, increases in agricultural productivity are considered here to be a necessary condition for the success of development strategy that bases itself on comparative advantage considerations".

According to James, Naya and Meier (1987) [11],

"The Asian countries that neglected agriculture in the pursuit of industrialisation through import substitution in the 1950s experienced an overall showdown in economic growth in the 1960s. India and the Philippines both adopted import substitution industrialisation strategies heavily biased against agriculture in the 1950s. The annual average real rate of economic growth in India was 3.6 percent during the first five year plan and 4.0 percent during second plan. Subsequently real growth fell during the third plan to an average annual rate of only 2.2 percent. In the Philippines the decline was from 8 percent in the first half of 1950s to less than 5 percent from 1956 to 1970. In contrast, Taiwan, which emphasised agriculture in the 1950s, experienced an acceleration of growth from 7.6 percent in the 1950s to nearly 10 percent in the 1960s. Taiwan and Korea adopted a `unimodal' strategy aimed at gradual modernisation of the entire rural sector, over one that seeks to concentrate resources in a highly commercialised subsector (Bimodel). Taiwan's experience indicates that the unimodal strategy has considerable advantages in maximising agriculture's contribution to overall development, while simultaneously raising rural living standard. Korea and the ASEAN-4 countries enjoyed generally greater growth of agricultural production (including food production) than the countries of South Asia. High rates of agricultural GDP growth are related to agricultural export performance. Agricultural exports grow fast in the countries with the high rates of agricultural growth. Higher rates of agricultural GDP in the ASEAN-4 than in the South Asian countries are related to differences in agricultural export performance".

In the present international context it thus seems that an investment strategy that favours food production in the context of peasant agriculture or of labour intensive business farms and that which induces industrialisation on the basis of agricultural growth is the most appropriate to ensure both sustained growth and improved welfare for the masses of the population, at least for countries which still have a large rural population with potentially large domestic markets and in which there already exists an industrial base with an established supply responsiveness.

The results of simulation experiment performed with a computable general equilibrium (CGE) model for Korea by Adelman [12] show that the ADLI strategy generates the same rate of industrialisation as does export led growth but leads to a higher rate of labour absorption, a better distribution of income, better balance of payment results, less poverty, a better distribution of income, and a higher rate of growth of per capita gross national product than export-led growth. These results are attained largely because of the greater increase in total factor productivity and larger reduction in unemployment with the ADLI strategy. Similarly, according to a paper presented at International conference of agricultural economists (1982), [13] CGE result for Egypt show that increasing the share of agriculture in total investment increases the GNP and raises the real income of all social classes.

6.3 Importance of Agricultural Exports: India

India was in transition in the 1970s, from inward looking policies to policies of relative outward orientation. She made sizeable efforts to diversify her exports both regionwise and commodity-wise. However, she aimed at exporting more and more of manufactured items. A whole range of subsidies and other incentives were offered to exports of manufactures and there had been a negative protection (through indirect taxation and quantitative restrictions in the form of licenses or quotas) in agriculture. The international competition was more in the field of manufactured exports which India choose to enter. India being a weak competitor in the world market, achieved a little success in manufactured exports at the cost of losing a major share of traditional exports in the world market. This led to increased imbalances in her Balance of Payments. The export promotion efforts in the manufacturing sector led to increased production of capital goods which in turn necessitated increased imports of raw materials and spare parts. Imports were liberalised. The liberalisation of imports without any parallel increase in capacity for their payments added to India's payments deficits. Remittances by migrant workers from overseas compensated to a large extent for the slow rise in export earnings. However, since 1979, the demand for foreign exchange earnings increased much faster than what the Indian economy was capable of earning. Huge finances were required to purchase mineral fuels as well as to maintain the pace of import liberalisation. Foreign aid, to which Indian economy was so habituated was not flowing in requisite quantum and was supplemented with borrowings from international commercial markets at the market rates of interest. As a result the debt servicing ratio increased. India, with a major proportion of its population below the poverty line hardly possesses the capacity to service debts at market rates of interest. The long-term solution for India thus is to strengthen its export base in order to gain increased export earnings. At this critical juncture of India's economic development, the feasibility of continuing to rely on manufacturing export-led growth as the major development dynamic is to be reassessed. There is urgency on the part of every sector to contribute more to exports "In an economy where demand chronically exceeds supply, each sector should generate export surplus commensurate with its share of national output and productive resources" (Sengupta 1986) [14]. Therefore, India's agricultural export potential needs to be rediscovered to augment exports.

The importance of agricultural and agro-based products in India's export trade can hardly be over-emphasised. Agricultural sector has been providing substantial support to export development right since fifties. The relative decline in its support as compared to manufacturing sector during seventies, eighties and mid of nineties can be revived in late nineties and 2000s as India is predominantly agricultural economy and is blessed with vast tracts of cultivable land endowed with favourable agroclimatic environment. Notwithstanding rapid industrial growth in the economy, the agricultural sector continues to play its predominant role in our economy. The linkage of the agricultural sector to the rest of the economy is so strong that the overall performance of the Indian economy is determined by its growth. Fall in agricultural output slows down not only the growth of per capita income but the subsequent growth of industrial production too. Agriculture is both a major source of purchasing power in the economy as well as the supplier of many raw materials needed for industrial production.

In the case of agricultural products, we have comparative advantage in basic production because of climatic and geographic conditions. Unlike other sectors of economic activity, this sector is characterised by the distinct feature that it requires precious little of imported inputs and functions for purposes of renewal of the exportable resources. Domestic resource cost (of each rupee) of foreign exchange earned is lower in agriculture than in manufactures (91 paise and 96 paise respectively). [Sengupta 1986] [15].

Export of agricultural produce can be of much potential benefit to the rural sector, as it removes the restriction on productive output imposed by the low domestic demand for food and provides the economic incentive to establish and improve the infrastructure in rural areas. The export opportunity allows the agricultural sector to expand productive capacity to the full extent. Factor incomes generated at home per rupee of output exported, directly and through backward linkages, amount to 71 paise for agriculture compared to 59 paise for manufactures and 69 paise for mining. (Sengupta 1986) [16].

Durable linkages can be built between agricultural export development and rural development with proper government management and intervention. Small and marginal farmers, the landless and the agricultural labourers can successfully participate in the production of agricultural products for export. Production of high value exportable crops would improve the economic conditions of small and marginal farmers. Living conditions of the tribal population and landless labourers may be improved with expansion of export markets for minor forest products.

Agricultural production is highly labour intensive and the gestation periods are short. Increased productivity would help in exporting at comparative prices.

Agricultural exports offer opportunities which can change the entire economics of cultivation for the growers. Prices are higher in export markets for certain commodities than in the home market (e.g. Basmati Rice, Sal oil, green vegetables, sheep/ goat, onions, potatoes, groundnuts, mushrooms and cut flowers). In the case of certain commodities, export markets are very vital. For example, the fruit and vegetable processing industry can be successful only with export market outlets; domestic demand alone cannot sustain it. If mutton and wool production have to be organised properly, it is necessary to ensure that the prices realised pay for the costs. This is possible only if part of the production is diverted to exports.

Basically, agricultural commodity exports require the least amount of cash assistance. Foreign exchange earning is net in most cases, as raw materials, capital requirement and components have not to be imported for agricultural production. No cash assistance is given to major export commodities. It is only in the case of selected processed agricultural items that some cash assistance is provided. In respect of some of the agricultural commodities like coffee, the government collects export duties.

Agricultural exports can be used gainfully for countertrade. Particularly food can be used for import of various other commodities in trading with some of the African and South Asian countries. India should try to exploit fully the potential of food, as the world demand for food would continue. Food is a major political weapon. Agricultural self-reliance, combined with agricultural exports, provides the necessary strength and wherewithal to stand stable and balanced without aligning with any superpower. Agricultural exports thus provide political strength to a nation. The terms of trade are slowly beginning to turn in favour of exports of foodgrains and other commercial crops.

6.4 Performance of Agricultural Exports: India

The performance of India's agricultural exports, in the present study, is evaluated in relation to their (i) comparative advantage, (ii) world agricultural exports, (iii) sectoral output, (iv) manufactured exports and (v) India's potential of horticultural and marine products.

Inter-country and inter-sectoral comparisons of composition of exports vis-à-vis RCA were worked out to evaluate the performance of India's agricultural exports in relation to their revealed comparative advantage. The share of India's agricultural exports in world agricultural exports was worked out to observe their performance as per world scenario and agricultural exports in relation to sectoral output and manufactured exports were worked out to see the performance of India's agricultural exports as per domestic variables.

6.4.1 India's Agricultural Exports and Comparative Advantage

The empirical results of Chapter 5 revealed that, consistent with the theory of shifting comparative advantage, the comparative advantage for Korea, Singapore, and ASEAN-4 countries shifted, with development, from agricultural products to labour intensive products and then to capital intensive products. In accordance with this shift in comparative advantage the composition of exports shifted from primary exports to labour intensive and then to capital intensive exports. In the case of India, however, except for a little shift towards labour intensive products by the beginning of 1980s, there was no significant shift in comparative advantage (disadvantage) for agricultural products (capital intensive products). The RCA index was above unity for agricultural and allied products till 1995 and there was marginal increase in this index from 1.35 in 1970 to 1.62 in 1975, 1.99 in 1980, 1.96 in 1985, and 1.91 in 1995. However, the composition of India's exports indicated a decline in share of agricultural exports from 40 percent in mid 1960s to 34 percent in mid 1970s to 28 percent in mid 1980s and further to 16 per cent in mid 1990s[Table 5.2 and 5.3].

Inter-sectoral comparisons of India's composition of exports vis-à-vis RCA (Table 5.2 and 5.3) indicates a higher percentage of exports of capital intensive products as compared to exports of agricultural products. By ranking the three broad product groups as per their RCA and composition of exports, agricultural products get the first rank as per RCA but second as per composition of exports. On the other hand, capital intensive products get the third rank as per RCA but first rank as per the composition of exports for the period 1970 to 1985. Though the share of labour intensive exports showed a little upward trend during 1980s, the share of agricultural exports in India's total exports showed a decline from 40 percent in mid-1960s to 17 percent by the beginning of 1980s and further to 16 per cent by mid of 1990s.

The inconsistency between the composition of exports and RCA ranking in the Indian case indicates poor performance of agricultural exports in India as per comparative advantage.

6.4.2 India's Share in World Agricultural Exports

India is a very large producer and consumer of agricultural commodities by world standards. It produces around ten percent of the world agricultural output, however, her share, at present (1994-95), in world exports of agricultural commodities is less than 1 per cent. The share of India's agricultural exports in world agricultural exports has been declining since the beginning of 1970s. From 2 per cent and above in mid 1960s (1966) it came down to more than 1 per cent in mid seventies, 1 per cent in mid 1980s and less than 1 per cent in late 80s and early 90s. There was marginal improvement in this share little above 1 per cent in 1995 (Table 6.1).

India is unable to retain its substantial share in respect of certain commodities like tea, spices, sugar etc. In tea e.g. India's once dominant position has been challenged by Sri Lanka and African producers. The share of India's tea exports in total world exports of tea has declined from 35 per cent in mid 1960s to 26 per cent in mid 1980s, 22 per cent in 1990, 20 per cent in 1991, 13 per cent in 1994 and 11 per cent in 1996. Similarly, the proportion of India's exports of spices in the total world exports of spices has declined from 25 percent in mid 1960s to 20 percent by beginning of 1970s, 14 percent by beginning of 1980s, 8 per cent by beginning of 1990s, and 9 per cent in 1994. The proportion of India's exports of fruits and vegetables has declined from around 2 percent in mid 1960s to little above 1 percent in 1970s, 1 percent in 1980s and only to 0.6 percent by beginning of 1990s. The fall in share of agricultural exports of India in total world exports is in spite of the fact that: (i) India enjoys a fairly large potential for agricultural horticultural and marine products due to large geographical area with varied climates, temperature and variety of soils accompanied by the largest coast lines in the world, (ii) demand for some of the higher value agricultural products such as meat, dairy items, marine products, fruits and vegetables and other processed and semi-processed products is increasing in western world leading to renewed preference for natural and fresh foods.

6.4.3 India's Agricultural Exports in Relation to Sectoral Output and Manufactured Exports

In an economy where demand chronically exceeds supply, each sector should generate export surplus commensurate with its share of national output and productive resources (Sengupta, 1986) [17].

Table 6.1 : Share of India's Agricultural Exports to World Agricultural Exports

Year	Agri Xs India/	Year	Agri Xs India/
1971	1.28	1984	1.02
1972	1.19	1985	1.03
1973	1.07	1986	1.03
1974	1.17	1987	0.92
1975	1.36	1988	0.75
1976	1.30	1989	0.88
1977	1.27	1990	0.95
1978	1.02	1991	0.93
1979	1.07	1992	0.91
1980	1.06	1993	0.99
1981	1.16	1994	0.83
1982	1.08	1995	1.24
1983	1.14		

Source : FAO Trade Year Books

However, India's agricultural exports account for only a small fraction of sectoral output. In 1994-95 31 percent of the GDP comes from the agricultural sector and only 19 percent of total exports consist of agricultural exports. The share of agricultural exports has been falling at a higher rate than the fall in the share of agriculture in GDP. In 1966-67 the share of agricultural exports in total exports i.e. 40.15 percent was more than the share of agricultural output in total output (36.7 percent). In 1970s also there was not much difference in these ratios i.e. share of agricultural exports to total exports was 35 percent in 1970-71 and 34 percent in 1975-76 and the share of agricultural output to total output was 40 percent in 1970-71 and 38 percent in 1975-76. However in 1980s though the share of agricultural output fell from 36 percent in 1980-81 to 31 percent in 1994-95 the share of agricultural exports in total exports fell from 31 percent in 1980-81 to 19 percent in 1994-95. A slight improvement in this share was visible in 1995-96 (19.2 per cent) [See Table 6.2].

The share of agricultural exports in manufactured exports showed continuous decline from 98 percent in 1966-67 to 63 percent in 1970-71, 55 percent in 1975-76, 47 percent in 1985-86 and only 27 percent in 1990-91, 23 percent in 1992-93 and further to 21 per cent in 1995-96 (see Table 6.2).

6.5 Reasons for Poor Performance of Agricultural Exports: India

Historical analysis of the countries under study shows that countries whose macro economic policies were conducive to agricultural exports and were based not on subsidies only but mainly on productivity and infrastructure, could show better performance in agricultural exports than countries whose policies discriminated against agricultural exports and production.

Macro economic policies on the one hand and lack of export surpluses on the other hand seem to be the two basic reasons for the

Table 6.2 : India's International Trade in Agricultural Commodities

Year	Share of Agriculture in GDP (percent)	Share of Agriculture and Allied Products in Total Exports (percent)	Share of Agricultural Exports in Manufactured Exports (percent)
1966-67	36.7	40.15	97.70
1970-71	39.7	34.90	63.07
1975-76	37.9	34.28	34.90
1980-81	36.3	30.65	54.90
1985-86	35.5	27.68	47.35
1986-87	31.7	27.48	43.83
1987-88	30.6	22.35	32.72
1988-89	32.4	18.40	24.91
1989-90	31.7	18.15	24.49
1990-91	31.6	18.50	26.61
1991-92	30.8	17.90	25.17
1992-93	32.1	16.50	23.16
1993-94	31.3	18.00	24.70
1994-95	30.61	18.66	21.20
1995-96	27.91	19.20	21.80

poor performance of agricultural exports in India. Inadequate export surpluses, in turn, seems to be due to (a) low rate of growth in agriculture, and (b) agricultural wastes. The low rate of growth in agriculture seems to be mainly due to (i) lack of infrastructure, and (ii) underutilisation of resources - land, irrigation and capital. The relationship is expressed in Table 6.3.

6.5.1 Macro Economic Policies

The inconsistency between the performance of agricultural exports and comparative advantage, in the case of India, seems to be mainly due to the macroeconomic policies. Macroeconomic policies may be categorised into: (a) "fundamentals" and (b) "selective" [18].

6.5.1.1 Fundamentals: Fundamental policies are set of these policies that contribute to the competitiveness of exports without intentionally discriminating between sectors and activities in an economy. These policies include (a) stable macroeconomic environment and a competitive real exchange rate, (b) financial deepening, (c) labour policies, (d) human resource development, and (e) the provision of high quality infrastructure.

It is evident from the successful experience of NIC-4 and ASEAN-4 that countries in which emphasis was on "fundamentals" as compared to selective policies, performed better with respect to international competitiveness and growth.

India, as compared to East Asian countries, lagged far behind in the long run aspects of fundamentals, particularly in the approach to human capital as well as infrastructure.

Macroeconomic environment: An important indicator of the stable macro economic environment of the East Asian countries is the low and stable inflation rate evident in each of these countries during their respective periods of export growth (with the exception of Korea in the late 1970s). India's average inflation rate for this

Table 6.3 : Reasons for Poor Performance of Agricultural Exports in India

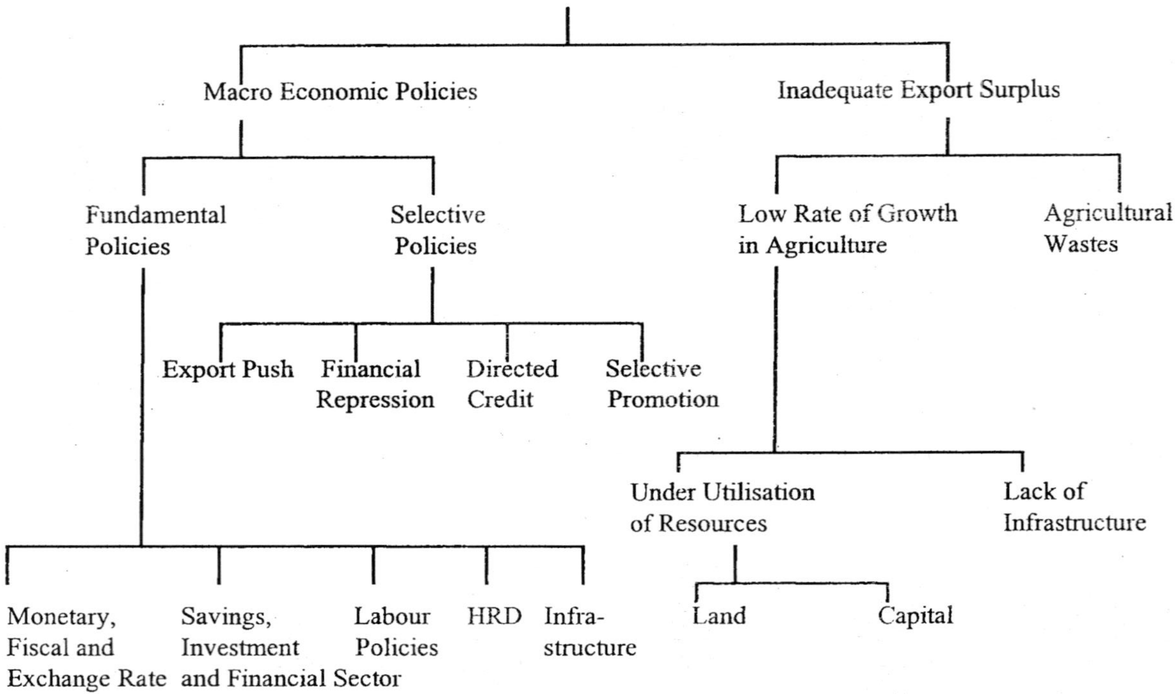

period is higher than almost all of the NIC-4 and ASEAN-4 (Table 6.4).

Perhaps the most important reason behind the low levels of inflation in the NIC-4 and ASEAN-4 countries has been the ability of the governments in these countries to keep the budget deficit within manageable limits. Almost all these countries have maintained fiscal discipline in varying degrees. On the other hand, Indian government has been particularly lax in maintaining fiscal discipline as compared to the East Asian countries. (Table 6.5). An important aspect of the fiscal consolidation in Korea, Malaysia and Thailand in the late 1980s is that most of the adjustments has occurred in the expenditure rather than the revenue side of the central government budget. In contrast, the deterioration in the Indian government's budgetary position has been caused mainly by the sharp increase in total expenditure since there is no evidence of a stagnation in total revenue collection (as percentage of GDP) during this period. Also, in contrast to the East Asian experience with fiscal policy, the Indian government's current account shows a deficit since the mid 1980s indicating that government has been borrowing to meet current expenditures. Since 1991, however, there has been a marginal reduction in the overall deficit (as percentage of GDP) as a consequence of the stabilisation programme that followed the macroeconomic crisis of that year.

Another indicator of macroeconomic stability is the degree of foreign indebtedness as measured by the ratio of external debt to export of goods and services. Though most of the NIC-4 and ASEAN-4 countries borrowed fairly heavily till the mid 1980s, with debt to GNP ratios comparable to other highly indebted economies (World Bank, 1993). However, given the higher levels of exports in these economies the ratio of total debt to exports was significantly lower than other developing countries (Table 6.6).

Human Resource Development :The link between human resource development (HRD) [or human capital] and economic growth is well recognised in the contemporary literature on the mechanics of economic development. Less clearly recognised

Table 6.4: Changes in Consumer Prices (percent per annum)

	Average (1981-90)	1990	1991	1992	1993	1994	1995	1996	1997
NIC-4									
HongKong	8.3	9.8	12.0	9.4	8.5	8.1	8.7	7.5	6.1
Korea	6.4	8.6	9.3	6.2	4.8	6.2	4.5	4.3	3.9
Singapore	2.5	3.4	3.4	2.3	2.4	3.8	3.2	2.4	2.8
Taipei, China	3.1	4.1	3.6	4.5	2.9	4.1	3.7	3.7	3.3
ASEAN-4									
Indonesia	8.5	7.8	9.4	7.6	9.6	8.5	9.4	7.5	8.0
Malaysia	3.6	2.7	4.4	4.7	3.6	3.7	3.4	3.7	4.0
Philippines	13.4	14.2	18.7	8.9	7.6	9.1	8.1	9.0	7.5
Thailand	4.4	5.9	5.7	4.1	3.4	5.1	5.8	5.5	5.5
India	8.8	8.9	13.9	11.8	6.4	10.2	9.5	9.0	8.0

Table 6.5: Overall Budget Surplus/Deficit of Central Government (percent of GDP)

	1981-85 (Average)	1985-90 (Average)	1990	1991	1992	1993	1994	1995
<u>NIC-4</u>								
HongKong	---	---	0.7	3.4	2.8	2.1	1.1	-0.2
Korea	-1.95	-0.28	-0.9	-1.9	-0.7	0.3	-0.6	-0.3
Singapore	2.29	5.48	2.7	4.7	5.5	4.4	3.6	5.1
Taipei, China	---	---	1.9	-2.2	-5.4	-3.9	-1.8	-1.7
<u>ASEAN-4</u>								
Indonesia	-1.19	-1.81	1.8	-0.7	-0.4	-0.4	0.2	-0.2
Malaysia	-9.95	-3.98	-3.0	-2.0	-0.8	0.2	2.4	0.3
Philippines	---	---	-3.5	-1.8	-1.2	-1.4	1.0	0.5
Thailand	-4.49	0.39	4.9	4.0	2.6	1.9	2.7	2.8
India	-6.80	-8.38	-8.3	-5.9	-5.7	-7.5	-6.5	-5.3

Table 6.6 : External Debt to Export Ratio (percent)

	Korea	Indonesia	Malaysia	Thailand	India
1971-75	156.4	168.7	39.1	66.7	314.1
1976-80	113.5	136.8	46.1	87.7	183.6
1981-85	130.6	132.0	96.0	142.8	244.0
1986-90	64.4	238.3	84.6	111.7	315.1
1990	53.8	259.3	54.7	53.5	456.1
1991	55.2	272.9	51.9	56.4	474.5
1992	57.6	259.1	49.0	58.8	460.2
1993	57.4	242.1	49.4	63.3	430.1
1994	56.8	241.0	38.3	56.8	394.8
1995	38.4	237.4	36.9	57.6	299.7

Source: Statistical Yearbook for Asia and the Pacific

perhaps is the link between human resource development and competitiveness. Formal education is the most important determinant of human resource development. Basic literacy may be considered to be sufficient at an early stage of industrial development. However, as the industrial sector becomes more technologically sophisticated, the technical orientation of the stock of human capital becomes increasingly important. Therefore, the process of human resource development, at this stage, requires a greater emphasis on science and engineering training as well as vocational and in-firm job training.

An important contributing factor to the superior performance of NIC-4 and ASEAN-4 economies in the area of exports is the provision of primary education to a large proportion of school-age population at a fairly early stage of each country's economic development.

India's HRD strategy is basically of a dualistic nature. India's performance in educational attainment at the primary and secondary levels has been much below that of the East Asian countries. An important aspect of India's HRD has been the technical orientation of its human capital-the percentage of scientists and engineers in total population and the proportion of tertiary level students in technical fields for India is comparable to that of other East Asian countries. However, the number of students in vocational education for India is significantly below several of the NIC-4 and ASEAN-4 countries. On the other hand, India has a surplus of technical manpower which often remains unemployed or leave the country. This reflects the "dualistic" nature of India's HRD strategy. An important indicator of HRD is the total public expenditure on education. The level of expenditure on education by the Indian Central Government is significantly below the level of expenditure by central governments of other East Asian countries. Further, out of this expenditure, the proportion on basic education is greater in East Asian countries as compared to India.

Infrastructure: Infrastructure is broadly defined to include public utilities such as power, telecommunications, water, public works such as roads and irrigation systems, and other transport sectors such as railways, urban transport, ports and airports. Certain infrastructural sectors (such as power, telecommunication and transport) are critical inputs in the manufacturing and production processes. An increase in both quantity and quality of the services of these sectors can increase output and productivity in the economy. India's performance in infrastructural quality and coverage has been creditable.

In the case of East Asian countries, one common aspect of these countries' policies towards infrastructure has been the priority they have given to the financing of infrastructure in their respective budgets. This has been particularly true for the 1970s and early 1980s. However, since the mid eighties, with most governments going through fiscal austerity programme, expenditure on infrastructure was curtailed or kept constant. Among the countries considered there is a wide variation in the government expenditure on infrastructure to GDP ratio. Similar to our finding in the case of HRD, there has been a weak commitment by the Central Government to the provision of infrastructure in India as indicated by the low level of Central Government expenditure on infrastructure as a percentage of GDP (as compared to East Asian countries).

The government should prioritise (as in Indonesia and Thailand) infrastructural expenditure (as in Indonesia and Thailand), first by putting basic infrastructure like roads, etc. in place and then paying more attention to the infrastructural sectors that are key to external trade such as ports and shipping. In the case of agricultural sector, the priority is ropeways, roads connecting production areas of agriculture to railway stations and ports.

An important indicator of approach to human capital and infrastructure is the total public expenditure on education and infrastructure. The level of expenditure on education by the Indian government is significantly below the level of such expenditure by governments of East Asian countries. Further, out of this expenditure, the proportion on basic education is greater in East

Asian countries as compared to India. Similar to our findings in the case of HRD, there has been a weak commitment by the Indian government to the provision of infrastructure in India as indicated by the low level of expenditure on infrastructure as a per cent of GDP. Thus, index of human capital and development of infrastructure is taken as expenditure by the government on education, fuel and energy, and on transport and communication as percentage of government's total expenditure (Table 6.7). This expenditure is above 20 per cent for all East Asian countries but for India it is below 5 per cent for the period 1986-1996 (Table 6.7).

6.5.1.2 Selective Policies: Selective policies are those which deliberately provide incentives to channelise resources into a specific sector or sectors of the economy. It includes as per EAM, four types of policies viz., (a) export push, (b) financial repression, (c) directed credit and (d) selective promotion. Selective policies were found to be successful only with reference to broad export export thrusts. Policies which tries to "pick winners" were generally found to be ineffective.

There were several similarities between the kinds of policy instruments used in the East Asian countries and India on the selective side. However, the successful use of discriminatory policy incentives is not so much an outcome of the incentives themselves, but rather of how the environment generates forces to ensure a high degree of compliance. The virtuous circle of export-led growth in East Asian countries was the result of the way in which export strategy was sequenced. India's problems with its export competitiveness and industrial efficiency stem not from the incentives it has used, but rather the sequencing of its sectoral emphasis. Even by 1973-74, the pattern of protection was heavily biased towards the capital goods sectors. Many of the reform steps taken since the beginning of 1991 were continuations and reinforcements along the lines of previous regime.

Table 6.7 : Expenditure by Central Governments on Basic Education and Infrastructure (percentage of total government expenditure)

	HongKong	Malaysia	Indonesia	Nepal	Singapore	Thailand	Rep. of Korea	India
1986	34.44	NA	17.29	21.78	28.76	27.22	23.53	4.01
1987	34.32	NA	19.38	23.09	24.47	27.60	23.96	4.57
1988	33.57	25.93	24.08	24.61	28.30	26.18	20.63	4.89
1989	20.14	26.14	22.23	23.92	23.85	28.13	20.11	5.02
1990	31.95	25.97	23.41	23.86	24.87	28.92	21.06	4.30
1991	33.12	25.62	25.09	19.02	27.33	31.60	21.10	3.99
1992	34.40	26.96	24.49	22.66	25.64	31.19	21.31	3.90
1993	33.36	28.03	25.56	26.68	27.25	35.13	27.44	3.59
1994	35.36	30.38	18.58	27.87	25.46	33.48	26.49	3.50
1995	33.73	30.25	17.20	26.00	22.12	35.80	28.33	3.70
1996	31.24	33.10	17.05	25.82	NA	35.87	27.86	3.70

It has been asserted that "India has an active policy of promoting the export of new manufactured engineering goods. She has neglected the traditional type of manufactured products". (Banerjee, p.26) [19].

We may state that India has neglected not only the traditional type of manufactured products but agricultural products too. Macro economic policies and trade policies were not conducive to expansion of agricultural exports in India while this was not the case in Korea, Taiwan and ASEAN-4.

"Major primary exports and traditional manufactured products were subject to export taxes in India. However, Korean incentive system was not biased against primary activities and within the primary sector, there was little discrimination against exports. In Taiwan also, there was little discrimination against primary activities taken together. In Malaysia, imposition of export taxes may have contributed to the fall in the share in the world export of rubber. This decline, however, has been partly offset by an improvement in Malaysia's position in the world tin market. On the other hand, Malaysia has experienced substantial increase in exports of some other primary commodities that are subject to practically no discrimination. The Philippines has made gains in sugar exports under preferential arrangements with the USA. It has also increased its share in the declining world exports of abaca and has expanded its timber exports" (Balassa, Bela 1971) [20].

The export promotion policies in India specially subsidies, have been directed towards sectors which did not possess comparative advantage. It needs to be stressed that quantitative restrictions in the form of licences or quotas have, more often than not, provided the means for regulating the volume of agricultural exports. Certain studies indicate that manufacturing has been receiving a positive protection and agriculture has been subjected to a negative protection (2.33 percent) through indirect taxation. The protection for agricultural sector has been low (NPC 0.90) as compared to manufacturing (NPC 1.4). It needs to be stressed that agricultural exports have been subject to quantitative restrictions in the form of licences or quotas till 1990. On the other hand planned import substitution resulting from stringent import controls to contain balance of payment deficits under the regime of overvalued exchange rate discriminated against sectors which did not have comparative advantage. Price benefits (due to government's support measures) and improved technology for the import-substituting crops resulted in not only increased yields in these crops as compared to the export crops, but also led to some shift of area from the export crops to the import substituting ones. The area under soybean crop increased from 2667 hectare in 1989-91 to around 5500 hectare in mid 1990s; and the area under cereal crops and pulses decreased from mid 1990s.

6.5.2 Rate of Growth in Agriculture

The macro economic policies i.e. cautious export promotion policies can lead to better performance of exports only if supplemented by an efficient supply management in the agricultural sector. Unless a sector (industry or agriculture) receiving protection is pushed rapidly to become internally competitive by not only producing more, but also by improving quality, it may not contribute significantly to improvement in competitiveness. "The effects of primary exports on industrial development depend to a considerable extent on output relationships" (Nicholls, William H., 1980) [21]. The main source of growth in agricultural exports of developing countries was the rise in domestic output. Rapidly rising demand in developing countries was met by rapidly rising domestic production.

Japan during the thirty years period between 1881-90 and 1911-20, increased its agricultural output by 77 percent, with the area under cultivation increasing by only 21 percent as compared with increased yields per acre of 46 percent. This means that in about thirty years, Japan was able to increase per capita food supplies by over 20 percent and the output per farm worker by 106 percent. Thus, the first and foremost requisite in exploiting India's agricultural export potential is to increase agricultural production significantly not only to meet the domestic requirements in full, but also to generate adequate export surpluses. Comparative analysis of growth of agricultural production by countries reveals that agricultural exports were growing fastest in the countries with the highest rates of agricultural growth (Table 6.8).

Table 6.8 indicates that growth rate of agriculture was lowest in the case of India as compared to Korea, Singapore, and ASEAN-4 countries in 1961-70, 1970-80 and 1980-91. The rate of growth during 1980-91 improved in the case of India but was still lower as compared to other countries, except Korea where the rate of growth in agricultural production as well as in agricultural exports slowed down. However, agricultural gains in India's agricultural production appear to have reflected to an unusual degree the effects of unusually favourable weather (during particular periods) while failing to show evidence of a persistent upward trend in output. The fact is, if 1987-88 and 1988-89 unusually good years are excluded when the growth rate was 13.2 percent and 6.00 percent respectively during the 11 year period 1980-91, the average growth rate in Indian agriculture will work out to be less than 2 percent for the period 1980-91. The period between 1991 and 1995 shows improvement in agricultural production, however, falling to show a persistent upward trend. The annual growth in agricultural production was negative for 1991 and 1995 but positive for 1992, 1993 and 1994 (Table 6.8).

6.5.3 Agricultural Wastes

A major cause of inadequate export surplus, along with low production rate in agriculture is the fact that the marketing of agricultural products tends to be characterised by high wastage. The wastage is estimated to range from 20 to 40 percent for fruits and vegetables and 20 to 30 percent for cereals [22]. This wastage, as per Swaminathan Committee Report, [23] is at different stages

Table 6.8 : Average Annual Growth Rate of Agricultural Production by Countries (percent)

Country/Period	1961-70	1970-80	1980-91	1991	1992	1993	1994	1995
India	1.8	2.5	3.9	-2.34	6.06	3.30	4.90	0.50
Korea	4.3	4.8	2.7	0.43	6.02	-2.89	1.60	3.70
Singapore	6.0	4.2	4.1	--	--	--	-0.18	1.15
Malaysia	5.8	5.2	5.2	0.01	4.75	4.34	-0.18	1.15
Philippines	3.6	4.9	5.4	---	---	---	---	---
Indonesia	2.3	3.4	4.8 (80-90)	---	6.26	1.66	0.56	4.18
Thailand	4.4	4.7	4.7	7.25	4.81	-1.35	5.34	2.50

Source: Handbook of International Trade and Development Statistics, United Nations

in the chain of picking, packaging, storage, transport and marketing due to poor production and marketing planning, insufficient storage, unsatisfactory transport structure and an inadequate organisational structure.

6.5.4 Inefficiency in Resource Management

Nature blessed India with land, 50 percent of which, against a global ratio of 10 percent, is suitable for cultivation. However, efficiency in the use of land has been low due to the fact that more than half of the estimated area of 328.8 million hectares, not presently under cultivation but potentially arable, has been lying waste due to the lack of economically viable technologies. Estimates for the area of wastelands in the country vary from 100 million hectares to 175 million hectares. Wasteland means land which is currently under-utilised and/or is deteriorating for lack of proper water and soil management or on account of natural causes. Wastelands are under ownership of semi government departments, communities, institutions and individuals. In a predominantly agricultural country like India where land is most important but limited natural resource, we can no longer afford to keep lands unproductive. There can be very little hope for this country so long as a third of its available land surface is almost completely non-productive and another one third is only partially productive.

Inefficiency in the use of capital is another major cause of low productivity in agriculture. The increase in output in several sectors has not been commensurate with the scale of investment undertaken. The emphasis on efficiency in the use of capital is doubly necessary at the present stage when resources available for public investment are well short of requirements. It would obviously be an unsound policy to undertake investments in new capacity in order to cover shortfalls arising from the poor utilisation of existing investments. Efficiency in the use of capital has often been low because of delays in implementation, low capacity utilisation and insufficient attention paid to efficient management and to the adoption of cost effective methods in all areas especially in irrigation, power transport and food processing units. With 1.5 million hectares available for aquaculture, we use less than 5 percent of it. In spite of being blessed with plenty of water by nature, only one third of our irrigated potential is fulfilled and even here our yields are extremely low when compared with the best in the world. India has 60 million hectares under irrigation, compared to just 47 million hectares in China and yet our foodgrains production is only 180 million tonnes whereas China produces two and a half times as much. With 7 percent and 17 percent share in World production of fruits and vegetables respectively, we process less than 1 percent of our total production commercially. In sharp contrast, Brazil processes 70 percent of its fruits and vegetable products, Malaysia 83 percent and the Philippines 78 percent. A survey of food processing units (including HPMC) at Shimla (1985) [24] revealed that capacity utilisation of these units was less than forty percent.

In short, we may state that the performance of India's agricultural exports has been quite poor (a) in comparison with other countries in the study; (b) in relation to sectoral output and industrial exports and (c) in terms of India's share in world agricultural export. The results of the present study seem to be consistent with the results of a World Bank study. According to this study, "India's agricultural exports have performed poorly, even worse than industrial exports". (World Bank, 1989) [25].

6.6 Policy Inferences for Agricultural Exports: India

Agricultural exports in India, are being treated as a peripheral and residual activity. "So far exports in the agricultural sector have essentially been in the nature of disposal of residuals. In any year, when surplus is available exports are undertaken, but in another year when surplus of that commodity is not available, exports are either stopped or slowed down. This, in turn, adversely affects the markets created with lot of effort and also our credibility as an exporting country. The policy of adhocism practised so far could not provide any consistency and stability of exports. It is thus imperative that exports of agricultural commodities will have to move from the stage of incidental activity to a planned deliberate effort" [26].

Export policies and strategy planning require, as a first step, identification of export sector and the commodities which would give us a comparative edge in terms of our resource endowments and capabilities. Having identified the commodities which could easily be developed for trade, the second foremost requisite for a developing country like ours is to create surpluses for exports. Thus two aspects of agricultural export strategy may be termed as - (a) What should we export ? i.e. the identification of commodities and (b) How to create surpluses for export, particularly in the identified commodities?

6.6.1 Identification of Commodities

To identify products having comparative advantage in agriculture, RCA index at three digit level was worked out for all the agricultural and allied products. The agricultural commodities having comparative advantage i.e. with RCA index above unity for throughout the period 1970 to 1995 were:

S.No	Group	Commodity
1.	031/034	Fish - fresh chilled or frozen
2.	051/057	Fruits fresh and nuts (excluding oil nuts)
3.	054	Vegetables fresh, chilled, frozen or simply preserved
4.	061	Sugar and Honey
5	071	Coffee and coffee substitutes
6	074	Tea and mate
7	075	Spices
8	081	Feeding stuff for animals
9	121	Tobacco unmanufactured
10	263	Cotton
11	264	Fibres including jute
12	291	Crude animal materials
13	292	Crude vegetable materials

The commodity groups having RCA index above unity since 1980 were:

S.No	Group	Commodity
1.	035	Fish - dried salted or in brine
2.	036	Smoked fish
3.	042	Rice in husk or husked but not further prepared
4.	058	Fruits preserved or fruit preparations
5	122	Tobacco manufactured
6	261	Silk

The RCA indices of the agricultural commodities having comparative advantage in India are presented in Table 6.9.

The performance of exports of the agricultural commodities having comparative advantage in India (the identified

Table 6.9 : Commodities Having Comparative Advantage i.e. Commodities with RCA Index above Unity 1980-96

Comm. gr → Year ↓	034	042	057	054	061	071	074	075	081	121	263	264	291	292	035	036	058	122	261
1970	3.52	-	4.78	1.90	1.56	1.50	51.54	31.37	-	6.14	1.73	7.90	5.52	6.83	-	-	-	-	-
1971	3.89	0.60	5.24	1.63	3.18	1.77	52.96	28.19	4.80	8.06	1.71	5.05	5.43	7.95	-	-	-	-	-
1972	4.73	0.61	4.82	0.81	0.88	1.72	54.75	22.62	5.03	8.70	2.09	17.01	5.61	6.54	-	-	-	-	-
1973	5.15	0.58	5.16	0.63	1.39	2.65	54.29	27.74	8.88	9.92	1.72	4.69	4.96	7.01	-	-	-	-	22.31
1974	5.27	2.36	6.78	0.83	5.50	2.96	63.35	36.59	7.34	10.40	2.01	22.86	10.47	10.64	-	-	-	-	31.61
1975	6.55	1.21	5.15	1.04	9.96	3.17	62.65	26.63	5.33	10.08	1.43	22.72	7.35	7.58	-	-	-	-	-
1976	6.95	1.27	4.69	1.62	6.98	2.31	58.38	24.03	7.18	7.43	3.07	3.98	5.95	6.41	-	-	-	-	-
1977	5.76	0.47	5.41	0.90	1.57	2.99	56.74	35.49	5.96	9.05	0.20	2.68	7.69	5.54	-	-	-	-	-
1978	N.A.	4.11	-	0.78	2.51	3.40	55.95	37.07	3.40	8.33	0.28	0	7.43	6.20	-	-	-	-	-
1979	N.A.	7.66	-	0.85	4.57	3.44	54.45	47.49	5.81	8.73	3.69	0	7.80	6.05	-	-	-	-	-
1980	N.A.	10.17	4.83	1.21	0.80	5.46	72.75	46.86	4.19	11.62	6.87	0	0.84	7.62	-	-	1.09	1.35	-
1981	N.A.	17.67	5.28	1.23	0.88	5.13	70.59	27.56	4.01	14.14	3.49	0	6.4	9.71	-	17.45	1.38	1.85	-
1982	N.A.	16.09	4.07	1.47	1.61	4.34	57.27	19.27	3.9	11.35	4.20	-	5.91	6.42	-	15.87	2.63	1.90	-
1983	1.35	6.20	2.86	1.56	2.66	2.51	62.72	21.02	2.85	6.32	5.21	1.17	4.39	4.92	1.08	1.56	1.38	1.04	20.11
1984	0.71	8.30	3.07	1.79	0.52	2.31	56.04	29.78	2.33	5.35	1.63	0.98	4.78	6.76	2.05	11.19	1.82	1.21	7.81
1985	0.84	12.54	3.91	1.70	0.30	3.85	68.57	45.02	2.96	5.81	2.41	0	5.05	6.64	1.38	13.21	1.03	1.45	1.53
1986	0.71	11.96	5.27	1.61	-	2.75	56.49	31.45	3.20	6.16	9.13	5.06	5.06	4.55	0.88	12.82	1.56	-	-
1987	-	18.07	4.03	1.03	-	3.15	53.17	25.88	2.54	4.20	3.50	-	4.12	4.93	0.47	9.95	0.72	-	-
1988	0.41	17.03	3.09	1.15	0.14	3.55	51.39	22.28	3.51	3.44	0.37	6.34	2.89	3.62	0.32	9.25	0.83	0.43	0.79
1989	0.48	13.92	2.74	1.03	0.36	2.92	44.73	19.14	4.12	3.51	0.94	3.28	2.57	3.16	0.83	7.51	1.39	0.39	1.70
1990	0.71	12.05	2.82	0.83	0.32	3.24	41.76	14.16	4.07	3.89	9.60	8.79	2.91	2.82	0.20	7.84	0.54	0.57	0.21

1991	0.74	13.79	2.97	0.92	1.19	3.08	38.57	16.99	4.40	4.22	2.87	1.95	2.67	2.64	0.20	8.10	0.48	0.36	0.22
1992	1.03	13.78	2.83	0.78	2.31	3.41	28.76	13.74	5.86	4.20	1.56	4.63	2.75	2.43	0.14	8.11	0.53	0.52	0.18
1993	1.12	14.20	3.09	0.87	0.91	3.65	22.88	16.68	6.90	3.87	5.23	1.32	2.62	2.39	0.14	8.87	0.58	0.33	0.33
1994	1.57	9.93	3.01	0.91	0.30	3.88	21.64	17.20	6.53	1.86	0.58	0.90	2.10	2.27	0.24	9.23	0.57	0.21	1.13
1995	1.35	29.91	2.58	0.99	1.71	4.44	25.35	14.70	5.43	3.46	0.61	0.56	1.89	2.86	0.21	7.09	0.58	0.17	0.56
1996	--	18.63	2.73	1.05	4.01	4.17	17.23	17.36	5.78	2.07	--	--	2.51	3.67	--	6.56	0.48	--	--

Source : Year Book of International Trade Statistics, U.N. (Various Issues).

commodities) however, has not been very encouraging. The share of these commodity groups in total world exports of respective commodity groups was much less during 1980 to 1995 as compared to 1970s (see Table 6.9). In tea, for example, the share of India's exports in total world exports has declined from 35 percent in mid 1960s to 25 percent by the end of 1980s and to 11 percent by 1996. Similarly, the proportion of India's exports of spices in total world exports of spices has declined from 25 percent in mid 1960s to 10 percent by the end of 1980s and 9 percent in 1995. The proportion of India's exports of fruits and vegetables in world exports has declined from above 2 percent in mid 1960s to less than 1 percent by end of 1980s and 0.6 percent by 1995.

The performance of exports of the identified commodities at micro level (i.e. at 3 digit level) is evaluated by comparing the revealed comparative advantage (RCA) of these commodities with the relative share of India in world exports of respective commodity groups. A detailed analysis of RCA indices of the identified commodity groups (Table 6.9) and percentage of India's exports of these commodities in world exports of respective commodity groups (Table 6.10) reveal the following :

In case of fruits and nuts (057), the RCA index varied between 3 and 6 throughout the period. The percentage in the world exports varied between 2 and 3 till 1981 and since 1982 it started declining and was little around 1.5 from 1982 to 1995 (except for 1986 when it was 2.45).

For vegetables fresh, chilled, frozen or simply preserved (054), RCA index was about 1 throughout the period (except 1979 and 1992 when it was 0.8) but the percentage in world exports was though more than 1 in 1970 varying between 1 and 0.5 till the end of 1980s and was insignificant i.e. below 0.5 percent from 1990 to 1992. A marginal improvement was visible from 1993 to 1995 when this share rose to a little above 0.5.

For fruit preserved or fruit preparations (058) the RCA index as well as the percentage of India's exports in world exports were nil before 1980. During 1980s the RCA Index was below unity and the

Table 6.10 : Percentage of India's Exports in World Exports of Respective Commodity Groups

YEAR	COMMODITY GROUP																
	034	057	054	071	074	075	081	121	061	263	264	291	292	035	036	042	058
1970	2.29	3.10	1.23	0.97	33.44	20.35	-	3.98	1.03	1.12	5.13	3.58	4.43	0.18	0	0	-
1971	2.28	-	-	-	-	16.51	2.81	4.72	-	1.00	-	3.18	4.66	0	0	0	-
1972	2.75	2.80	0.47	1.00	31.82	13.14	2.92	5.06	0.51	1.21	9.88	3.26	3.18	0	0	0.35	-
1973	2.65	2.65	0	1.36	27.87	14.24	4.56	5.09	0.71	0.88	2.41	2.55	3.60	0	0	0.29	-
1974	2.46	3.16	0	1.38	29.52	17.05	3.42	4.85	0.03	0.93	0.12	4.88	4.96	0	0	0.10	-
1975	3.28	2.57	0.52	1.59	31.32	13.32	2.67	5.04	4.98	0.72	11.36	3.67	3.79	0	0	0.60	-
1976	3.75	2.53	0.87	1.24	31.50	12.97	3.87	4.01	3.77	1.65	2.15	3.21	3.45	0	0	0.68	-
1977	3.07	2.88	0.48	1.59	30.26	18.98	3.18	4.83	0.84	0.11	1.43	4.10	2.95	0.06	0.19	0.24	-
1978	-	-	0.37	1.62	26.62	17.68	1.62	3.97	1.20	1.34	-	3.54	2.95	0.06	0.23	1.96	-
1979	0	-	0.37	1.47	23.32	20.34	2.49	3.74	1.96	1.58	-	3.34	2.59	0.58	0.21	3.27	-
1980	0	1.81	0.45	2.05	27.33	17.60	1.57	4.37	0.30	2.58	-	0.32	2.86	0.06	0.19	3.82	0.41
1981	0	2.11	0.49	2.04	28.14	10.99	1.60	5.64	0.35	1.39	-	2.55	3.88	0.07	0.20	3.04	0.55
1982	0	1.88	0.68	2.00	26.44	8.90	1.80	5.24	0.74	1.94	-	2.73	2.96	0.06	0.24	7.42	1.21
1983	0.39	1.49	0.81	1.31	32.69	10.96	1.49	3.30	1.39	2.88	0.61	2.29	2.57	0.06	0.28	3.23	0.72
1984	0.20	1.60	0.93	1.20	29.18	15.51	1.21	2.79	0.27	0.85	0.51	2.49	3.52	0.05	0.26	4.32	0.62

1985	0.25	1.79	0.78	1.76	31.31	20.56	1.35	2.66	0.13	1.10	-	2.31	3.03	0.05	0.26	5.72	0.93
1986	0.33	2.45	0.75	1.28	26.25	14.62	1.49	2.86	-	4.24	-	2.35	2.12	0.41	5.96	5.56	0.72
1987	0.21	1.97	0.50	1.53	25.92	12.62	1.24	2.05	-	1.71	-	2.01	2.41	0.21	4.85	8.81	0.35
1988	0.41	1.52	0.56	1.74	25.20	10.93	1.72	1.69	0.07	0.18	3.11	1.42	1.77	0.16	4.54	8.34	0.41
1989	0.48	1.53	0.58	1.65	25.21	10.79	2.32	1.98	0.21	0.81	1.85	1.45	1.78	0.46	4.23	7.79	0.78
1990	0.71	1.49	0.44	1.71	22.06	7.72	2.15	2.05	0.17	--	4.62	1.54	1.49	0.10	4.14	6.37	0.28
1991	0.74	1.55	0.48	1.61	20.08	8.85	2.29	2.19	0.62	1.50	1.02	1.39	1.38	0.10	4.22	7.18	0.25
1992	1.04	1.58	0.44	1.91	16.08	7.68	3.28	2.35	1.30	0.87	2.59	1.54	1.36	0.07	4.54	7.71	0.30
1993	1.13	1.86	0.53	2.20	13.79	10.05	4.16	2.33	0.55	3.15	0.80	1.58	1.44	0.08	5.35	8.56	0.35
1994	1.58	1.88	0.57	2.42	13.48	10.71	4.06	1.16	0.19	0.36	0.56	1.31	1.41	0.12	5.75	6.19	0.36
1995	1.36	1.63	0.63	2.81	16.03	9.30	3.44	2.19	1.08	0.45	0.41	1.20	1.81	0.16	4.49	18.92	0.37
1996	–	7.76	0.68	2.68	11.09	11.17	3.72	1.33	2.58	–	1.91	1.62	2.36	–	4.22	11.99	0.31

Source : Year book of International Trade Statistics, U.N. Various Issues.

percentage in world exports was 0.4 in 1980, 1.2 in 82, 0.7 in 1983, 0.6 in 1984 and 0.9 in 1985, was insignificant to 0.3 from 1985 to 1995.

Exports show large amount of variations in the case of sugar and honey (061). In 1970 the share of the sugar and honey exports was above 1 percent, below 1 percent from 1971 to 1974, increased to 5 percent in 1974, decreased to 4 percent in 1975 and further to 0.8 percent in 1976. In 1978 and 1979 the share was 1.2 and 1.9 percent respectively and it fell to insignificant proportion i.e. 0.3 percent and 0.2 percent between 1980 and 1995 (except for 1992 and 1995 when it was above 1).

In case of coffee and coffee substitutes (071) a large variation throughout the period is noticed. The percentage of India's exports in world export's was less than 1 in 1970, varied between 1 and 2 during 1970-79, 2 and 3 in 1980-82 and again fell between 1 and 2 during 1983-85 and 1.2 in 1986 and 1.5 in 1987. There was slight improvement in this share in successive years till it reached to 2.8 in 1995.

In case of Tea and mate (074) the RCA index was highest, varying between 50 and 70 from 1970-88. There was a decline in this index from 1980 to 1995. The percentage of 074 in world exports was varying between 25 and 30 till 1988. There was persistent fall in this share from 1988 to 1995 when it reached 13 percent in 1994 with a slight revival in 1995 to 16 percent.

Percentage of this group in India's total exports was 16 in 1967, 10 in 1970 and varying between 5 and 6 in successive years till 1985 and fell to 4.6 in 1986 and 3.3 in 1987, 1.17 in 1994 and 1.09 in 1995.

Spices exports (075), having very high index of RCA (varying between 25 and 45 during 1970-85) showed variation in percentage of its exports in total world exports with a declining trend from 25 percent in mid 1960s to 20 percent in 1970 to 15 percent in 1984 and further to 12 percent in 1987. However, there was persistent fall in its RCA index from 1988 to 1995 when it was only 14.7 percent with percentage of exports falling to 9 percent in 1995.

Feeding stuff for animals (081) show variation between 3 to 8 during 1970-1995 in RCA index with declining trend in 1980s and a little revival from 1990 to 1995. The percentage of this group in world exports showed declining trend from 3 percent in beginning of 1970s to 2 percent in late 1970s and 1 percent in 1980s. There was an upward revival in this ratio to 3 percent by 1995.

For tobacco unmanufactured (121) RCA index showed variation from 6 to 14 during 1970s with a decline to 5 by mid of 1980s and to 3 by mid of 1990s. The percentage of this group in world exports hovered around 3 from the beginning of 1970s to beginning of 1980s but fell to less than 2 percent by 1988-89 and was around 2 from 1990 to 1995.

Cotton exports (263) also showed variation with RCA index varying between 1 and 6 from 1970 to 1986, 1987 with variation from)0.3 to 9.6 from 1988 to 1995 and percentage of India's exports of this group in total world exports varying between 1 and 2 till 1987 and to an insignificant proportion of 0.4 from 1990 to 1995 (except for 1993 when it was 3 percent).

Percentage of India's exports in world exports of crude animal materials (291) showed decline in 1980s as compared to 1970s.

Similarly, crude vegetable material (292) showed variation in RCA index as well as in percentage of India's exports in world exports for 1970 to 1995.

In short, agricultural commodities, having comparative advantage in India, have been losing their share in the world market. For example, India's once dominant position in tea has been challenged by Sri Lankan and African producers. The falling share of agricultural raw materials (i.e. 081, 121, 263, 264, 291 and 292) may be attributed to increase in domestic demands for raw materials due to increase in value added product exports and possibly slow growth prospects in world markets. The prospects of high growth in exports of tropical beverages (i.e. 071 and 074) may be limited due to slow growth prospects in world markets and of food exports such as cereals, live stock products and sugar, due to subsidised exports of developed countries. [This does not imply that India should not maintain its share of these traditional exports.] The fall in share of horticultural exports of India in total world exports, however, is in spite of the fact that - (a) India enjoys a fairly large potential for horticultural and marine products due to large geographical area with varied climates, temperature and variety of soils accompanied by the largest coast lines in the world, and (b) world demand for some of the higher value agricultural products such as meat, dairy items, marine products, fruits and vegetables and other processed and semiprocessed products is increasing in western world leading to renewed preference for natural and fresh foods. The potential that exists for India to increase her share in the world markets of horticultural products may be indicated by the fact that India's share in production for fruits and vegetables as a group is around 11 percent of the world total yet her share of world exports of this group is less than 1 percent. India is the third largest producer of fruits after Brazil and USA. However, USA and Brazil are the main exporters of fruits in the world market whereas India's position is marginal.

The declining share of India's exports of agricultural commodities (having comparative advantage) in world market may indicate that these commodities received less attention in realising their export potential. Thus, we may state that apart from maximising the exports of traditional commodities and maintaining their share in world exports such as spices, tea and coffee; importance of promoting non-traditional exports with high growth prospects in world markets and high production potential at home such as horticulture, apiculture, fisheries and sericulture can hardly be overemphasised. The exports of non-traditional commodities like fruits, vegetables, flowers and honey require greater attention in realising the country's potential of agricultural exports.

6.6.2 Measures to Create Export Surpluses

The second foremost requisite for a developing country like ours is to create surpluses for export. The effects of primary exports on industrial development depend to a considerable extent on output relationships. "Agriculture could be exploited for general development and trade only if it first produced a surplus, which was there for exploitation" [27].

The present level of production after meeting vast internal demand leaves meagre export surpluses. In a developing economy like ours, with its huge size and population, the levels of consumption within the country are important considerations. However, looking at exports only as a means of disposing of the available surpluses cannot form the basis of a healthy export policy. Obviously, it would be necessary to increase agricultural production significantly not only to meet the domestic requirements in full, but also to generate adequate export surpluses. The main source of growth in agricultural exports of developing countries was the rise in domestic output. Rapidly rising demand in developing countries was met by rapidly rising domestic production, the proportion of output that was exported was low. This implies that significant increase in agricultural production may be the foremost requisite in exploiting India's export potential. The basic push has to come through (a) yield improvement, (b) development of adequate infrastructure, (c) reduction in waste, and (d) efficiency in resource management.

6.6.2.1 Yield Improvement: India's potential for increasing output through yield improvement is significant as the untapped yield potential is quite high in most of the farming systems, in the country. India is first largest producer of groundnut in the world but 72nd in yield, second largest producer in rice and wheat but 54th and 42nd in yield. In fruits and vegetables, a large number of agricultural crops (almost all the crops) in this country suffer from low yields as compared to (a) world average, and (b) Asian countries average (see Table 6.11).

Table 6.11 : Average Yield (Kg/HA) of Selected Agricultural Crops in India, World, Asian Countries

S. No.	Product	Year	World Average	Asian Countries Average	India
1.	Cereals	1996	2908	3164	2161
		1997	2967	3158	2245
		1998	2970	3162	2207
2.	Pulses	1996	780	694	543
		1997	807	704	584
		1998	837	727	584
3.	Soybean	1996	2119	1423	995
		1997	2137	1405	955
		1998	2240	1353	961
4.	Cabbage	1996	23759	25181	18182
		1997	23758	24874	18261
		1998	24093	25066	18261
5.	Tomatoes	1996	28355	25209	15143
		1997	27494	25170	15143
		1998	28343	25898	15143
6.	Cucumber	1996	16873	17688	6686
		1997	17111	17739	6705
		1998	17007	17568	6705

7.	Onions (Dry)	1996	16635	16085	10074
		1997	16844	16263	10672
		1998	17269	16839	10672
8.	Garlic	1996	10867	11614	3984
		1997	11031	11941	3996
		1998	11075	11946	3966
9.	Cauliflower	1996	18243	19300	16071
		1997	18678	19829	17421
		1998	18799	19822	17241
10.	Beans (Green)	1996	6609	6420	2703
		1997	6853	6652	2703
		1998	6774	6716	2703
11.	Peas (Green)	1996	8174	8870	4120
		1997	8302	9156	4286
		1998	8256	9140	4286
12.	Carrots	1996	22367	22186	14545
		1997	22865	21425	14468
		1998	23333	21747	14468
13.	Watermelon	1996	18712	21255	31006
		1997	19012	21723	13158
		1998	19052	21917	13158

Source: Production Yearbook, FAO, 1998.

If this is corrected, per hectare yields would increase to take care not only of the increases in domestic consumption but at the same time export surpluses will be generated. As per data in Table 6.10 authenticated by the extensive experimental work done by Pusa Institute for Agricultural Research, a 100 to 200 percent increase in yields for vegetables appears conceivable. Improvements in yields may be achieved by better use of fertilisers, irrigation facilities, high yielding varieties of seeds and mechanisation.

6.6.2.2 Development of Human Capital and Basic Infrastructure: Production increase will prove to be effective i.e. cost disadvantage will be reduced only if backed by adequate public investment in human capital and infrastructure. Any growth process based on entrepreneurial response in rural areas will depend very heavily on social investment in primary education and health. No self-sustaining growth is possible if economic policies neglect their crucial sectors. Infrastructure in the form of ports, railways and roads represent important inputs for primary exports and eventually contribute to the development of industrial activities. Empirical studies show that the output of individual crops is quite responsive to changes in the relative price of one crop with respect to other. But the supply response of overall agricultural output to a proportional increase in the prices of all crops is actually low. In contrast, the elasticities with respect to technology (such as availability of improved varieties) and infrastructure (such as irrigation and roads) variables, which are the outcome of public investment are typically high. An area endowed with a good network of irrigation, roads and power will yield high returns on agricultural investments.

Education, health and basic infrastructure (energy, transport, communication, irrigation) were listed as priorities in the approach to the Eighth Five Year Plan. It was stressed in the approach that sustained growth of the economy depends on adequate availability of basic infrastructural facilities and human development. However, the Eighth Plan had fallen significantly short not only of

Table 6.12: Performance of Infrastructure Sector in India

S.No.		VII Plan	VIII Plan*
1.	Irrigation Potential (million ha.)	11.3	10.6*
2.	National Highway Network	1760	609*
3.	Power Generation Capacity	21401	17667*
4.	Education	3.51	3.89*

* Anticipated
Source: VIII Five Year Plan, 1992-97

the targets but also of Seventh Plan, in these areas (Table 6.12). The primary reasons for this are: (a) efforts at containing the fiscal deficit of the government have fallen disproportionately on public investment, particularly in those sectors which are more dependant on budgetary support. The share of capital expenditure in total expenditure of government declined sharply from 30 per cent at the beginning of Eighth Plan to 24 percent at the end of Plan. With a considerably lower level of public investment (33 percent of total investment in the economy) in the Ninth Plan as compared to 43 percent (targeted) and 36.5 percent (anticipated) for the Eighth Plan, it is doubtful that Government of India would be able to provide adequate infrastructure and productive services for the agricultural sector.

Thus, a necessary condition for improving competitiveness of agrucultural exports is provision of adequate support infrastructure (e.g. roads, telecommunications, port facilities) and a network of producer services (e.g. transport, packaging, cold storage, credit and repair services) by the government by increasing public investment in these sectors. Efforts at containing the fiscal deficit by the government should fall on unproductive expenditures than on priority sectors.

The experience of Japan shows that appropriate expenditure by government (on agricultural research, extension, credit and roads) can have spectacular effects on the output of farmers and that agriculture instead of acting as a brake on the rest of the economy, can be turned into a leader, generating demand for other sectors, besides providing them with requisite capital. According to a Federation of Indian Chambers of Commerce and Industries (FICCI) Report, one of the major factors for better performance of all the four Asian NICs was development of adequate infrastructure. All the four Asian NICs placed greater emphasis on the development of infrastructural base, such as transport network, roads, railways, ports, and shipping, communication, power and the like.

The Government of India is also keen to develop infrastructural facilities by providing various types of economic services for promoting the overall development climate in the country. However, at present, as stated earlier, we still lag behind other countries in this respect, specially in case of agricultural sector. The main requisite at present, along with the development of ports, railways and roads in urban areas is to connect production areas of agricultural sector in villages and hills (through roads, railways and ropeways) to urban areas and ports.

6.6.2.3 Reduction in Wastes: The marketing of agricultural products tends to be characterised by high wastage. Around 30 percent of agricultural products is estimated to be wasted due to lack of adequate processing facilities, absence of adequate infrastructure for storage, rapid transportation etc. [28]. This wastage can be reduced by a proper planning at the production and post harvest phases, organising trade on a systematic basis and improvement in availability of producer (economic) services. Agricultural and agro-processing activities may require producer (economic) services like transport, repairs and storage. The success story of rural growth in Punjab and Surat District in Gujarat suggests that productivity growth is associated with two distinct but related sub-processes viz., (a) ever increasing diversification of the economy and (b) rural transformation through special diffusion of new activities. The availability of producer services plays a big role in both of them. Thus, if 30 percent more is added to the yield improvement, which may be realised by eliminating the present wastes, the increase in production may come out to be still higher.

According to the estimates by Panchmukhi (1983) [29], "the overall productivity of expenditure on economic services in terms of their contribution - either directly or indirectly to the change in NDP is quite impressive. For every additional rupee spent on economic services of agriculture and allied services, NDP would show an increase of about Rs.13.24. Similarly the productivity of revenue expenditure on industry and mineral services is Rs. 3.84. The average elasticities of NDP with respect to expenditure on economic services works out at 0.11 and 0.08 respectively for agricultural and industry services. Both the marginal and average productivities of expenditure on economic services have been increasing in recent years". However, the government expenditure on economic services specially in the case of agricultural sector is low compared to other countries. It was 18.01 percent in 1990 and 18.58 percent in 1995.

6.6.2.4 Efficiency in Resource Management: A major cause of low productivity in agriculture is the inefficiency in the use of resources - land and capital.

Efficiency in the use of capital has often been low in the past because of delays in implementation and insufficient attention paid to efficient management. Thus, improvements in capacity utilisation and efficient project implementation in all areas, especially in irrigation, power, transport and food processing units are essential for achieving increases in productivity of capital in the agricultural sector.

Efficiency in the use of land has been low due to the fact that more than half of the estimated land area of 328.8 million hectares remains unutilised in India. In a predominantly agrarian country, where land is the most important but limited natural resource, it is imprudent to allow huge land mass to lie waste. Thus it becomes imperative to make the best possible use of our wasteland resources through scientific land management. "Arresting the declining productivity of the national resource base will hinge on improving the management of property held in common [30] (World Bank Country Study 1989). In Indian climatic conditions, most of these wastelands are capable of sustaining some vegetation if suitable techniques are adopted, necessary inputs provided and minimal discipline in tending the crops observed. Such vegetal covers would help the improvement in environment, ensure greater soil and moisture conservation and produce a kind of renewable wealth (including foreign exchange) so much needed in India today. Considering the poor production of land, nature of risk involved, employment potential, demand for the outputs and overall economic gains, afforestation seems to be the appropriate technology for development of wastelands. Much of the theoretical work has been done and several initiatives taken, following the creation of National Wasteland Development Board (NWDB). The NWDB was established in 1985 with the objective of bringing 5 million hectares of land every year under fuel, wood and fodder plantations. The afforestation of 5 million hectares, however, will need 10 billion saplings and would cost more than Rs. 25,000 million annually. The need is to discover, explore and to experiment not only in how to raise nurseries but also whole plantations. Thus, it is necessary to adopt commercial approach for the wasteland development programme.

Utilisation of wastelands for horticulture along with afforestation seems to be efficient use of land resources as (a) The wasteland which is unsuitable for field crops can profitably be utilised for horticultural crops, (b) horticulture pursuits are labour intensive, provide high income per unit of land and help in maintaining ecological balance, and (c) Horticulture, in addition to supplementing the national foodgrid by providing nutritive food in the form of fresh and processed fruits and vegetables, can be an important source of foreign exchange earnings. Certain species of fruit trees, vegetables and medicinal plants, which can be grown on wastelands are an important source of agricultural exports. Thus afforestation supplemented with horticulture, floriculture, appiculture and food processing in urban areas can assure substantial amount of agricultural exports and afforestation supplemented with other related vocations such as bee-keeping, livestock raising, charcoal making, carpentary, paper and pulp production can assure year round employment in rural areas. Efforts need to be made first, to carry out micro level surveys for systematic identification of the extent, location and nature of wastelands suitable for plantation work and second, define specific development projects (for horticulture, floriculture and appiculture etc.) that necessarily will entail site specific work. A processing plant with a marketing system should be designed along with this development plant to link it with a strong marketing system.

6.7 Conclusion

At present, India confronts a difficult situation in its international competitive position. The staggering trade deficits of the past several years and the borrowing needed to finance them obviously can not be sustained for long. The long term solution for India, thus, is to strengthen its export base in order to gain increased export earnings. The feasibility of continuing to rely on manufacturing export led growth as a major development dynamic is to be reassessed at this critical juncture of India's economic development. The export potential of the agricultural sector needs to be rediscovered to augment exports in India. The performance of India's agricultural exports has, however, been quite poor (a) in comparison with other countries in the study, (b) in industrial exports, and (c) in world agricultural exports. The share of agricultural exports in India's total exports showed decline from 40 percent in mid 1960s to 17 percent by beginning of 1990s. Similarly, the share of India's agricultural exports in world agricultural exports declined from 1.5 percent in 1960s to 0.8 percent by end of 1980s. The explanation for this paradoxical situation lies in the slow growth of agricultural production due to low yields, wastes and inefficiency in resource management. Thus the basic push to agricultural exports in India, along with policy initiatives, has to come through supply management in the agricultural sector by (a) yield improvement, development of human capital and basic infrastructure (b) reduction in wastes, and (c) efficiency in resource management accompanied by conscious export promotion efforts.

Notes

1. The analysis of the present chapter as mentioned in the scope of study in chapter VI, is limited to a case study of India due to the vastness of the subject and limitation of comparative data.

2. Adelman, and Irma (1984) - World Development (1984), p.219.

3. Ibid.

4. According to Bela Balassa "Though in the long run, learning to export manufactures is crucial for development. A nation that cannot export manufactures can hardly be considered fully developed since its supply capabilities will be very narrow. However, it is difficult to achieve increase in manufactured exports on any significant scale until industrial development reaches a fairly advanced level. The reasons are: (a) in its early stages, industrial development is strait jacketed by the small domestic market and is forced to focus on `Make-or-buy decisions for that market'. (b) The competition is fierce in export oriented manufacturing and standards are rigorous. (c) Unlike primary exports, there is no significant rent from natural resources." (Balassa, Bela, World Bank Staff Working Paper, 53).

5. Adelman and Irma (1984), World Development (1984), p. 220.

6. Hans Singer, Neelamber Hatti and Rameshwar Tondon (1984). International Commodity Policy.

7. Ibid.

8. For linkage effects see Hans Singer, (1984) ; op.cit., p.27.

9. Adelman and Irma (1984), World Development (1984).

10. R. Banerjee (1975) - Export of Manufactures from India.

11. James W.E., Naya, Seiji and Meier, G.M. (1987) : Asian Development - A Historical Perspective.

12. Adelman and Irma (1984), World Development (1984).

13. Implications for Agricultural Policies in Developing Countries - International Conference of Agricultural Economists (1982).

14. Sengupta, S.K. (1986) : Export Prospects of Cereal, Fruits and Vegetables (FICCI Report, 1986).

15. Ibid.

16. Ibid.

17. Sengupta, S.K. (1986) : op.cit.

18. World Bank, 1993: The East Asian Miracle: Public Policy and Economic Growth.

19. Banerjee (1975) : Exports of Manufactures from India.

20. Balassa, Bela (1971) : Structure of Protection in Developing Countries.

21. Nicholls, William H. (1980) : "The Place of Agriculture in Economic Development"

22. Eight Five Year Plan.

23. Government of India (1981) : Major Recommendations of the Group on Perishable Agricultural Commodities: M.S. Swaminathan Report.

24. The survey was conducted by the students in 1985.

25. World Bank (1989) : World Bank Country Study : India 1989.

26. Ministry of Agriculture (1988).

27. Eiecher and Witt (1980) - Agriculture and Economic Development.

28. Eigth Five Year Plan

29. Panchmukhi, V.R. (1983) : Neglect of Economic Services - Economic Times, 9th March, 1983.

30. World Bank (1989) : World Bank Country Study : India.

Chapter 7

Implications of Trade Policy Reforms on India's Agricultural Exports

7.1 Introduction

The analysis of Chapter 5 revealed that the competitiveness of exports of South Asian countries may be improved by increasing the share of agricultural exports in the total exports of these countries. The analysis of Chapter 6 (a case study of India), however revealed that the performance of India's agricultural exports has been quite poor as per (a) India's sectoral output, and (b) world scenario. It was further inferred (Chapter 6) that the explanation for this paradoxical situation lies in (a) lack of policy initiatives in the field of agricultural exports, and (b) slow growth of agricultural production due to low yields, wastes and inefficiency in resource management. Thus, the basic push, as per the proposed strategy of agricultural exports (Chapter 6) may come through (a) conscious export promotion efforts i.e. policy initiatives in the agricultural sector, and (b) supply management by improvement in yields, reduction in wastes and efficiency in resource management.

The common belief, at present, is that trade related reforms introduced since July 1991, mainly during annual budgetary sessions,have begun removing major impediments by reversing India's past inward-looking policies and integrating India more fully into the world economy. Its already high profile and interest in the multilateral trading system has also been strengthened. The trade reforms, if supported by appropriate macro-economic policies, should lead to improved resource efficiency and help increase India's international competitiveness on both domestic and export markets. Macro-economic stabilization and structural policies, along with sector specific policies have an important bearing on exports and efficient growth of the economy. Sustained macro and micro economic reforms should also assist in stabilizing the Balance of Payments situation. The agricultural sector, which until now has retained self-sufficiency as a goal, could on balance, also be a major beneficiary of more liberal trading measures. The implications and consequences of trade policy reform, however, are contingent on the maintenance of an open, stable and predictable global trading regime that can successfully combat protectionist pressures in major export markets. The multilateral liberalization of trade in agriculture and the introduction of an international regime of discipline for domestic support to agriculture (in the Uruguay Round of GATT negotiations) would obviously influence the output of and trade in agricultural commodities in the world economy. This in turn is bound to have significant implications and consequences for the agricultural sector. The long-term impact on India's agriculture and agricultural exports would be determined by an interaction of the changes in national trade policy and international trade policy. Thus, the objective of this chapter is to analyse the impact of trade policy reforms in the national context and multilateral trade policy changes in the international context on India's agricultural exports.

Trade policy vis-a-vis the industrial sector has implications for the agricultural sector. The policy of import substitution associated with over valued exchange rate may discriminate not only against exports but also against the agricultural sector. Also the regime of export incentives confined mostly to manufactured exports may increase relative profitability of producing tradeable in the manufacturing sector as compared to the agricultural sector. This implicit bias against agriculture may lead to a misallocation of resources. Thus, in the following section (Section 7.2) an attempt is made to analyse as to what extent the trade policy of India till the Seventh Plan period was conducive to increasing agricultural exports. If not, was there any improvement in this regard due to the trade policy reforms introduced in 1991-92 and 1992-93 budgets and operationalized in the export-import policy (EXIM policy) for the eighth plan period ? However, as the external environment has also important implication for the agricultural sector, an attempt would be made in Section 7.3 to analyse the likely impact of agreement on agriculture under the Uruguay Round of GATT negotiations.

7.2 Trade Policy: India

7.2.1 Trade Policy till the Fifth Five Year Plan

The present trade policy has to be seen in the perspective that the concomitant desire for economic independence, with the inception of national planning, led to a development strategy of rapid industrialization in India. Industrialization and economic development became virtually synonymous for political leaders and economic planners. Insufficient attention was paid to agricultural development as a means to promote overall economic growth. Trade in agriculture was excluded from trade policy regime and it was implicitly assumed that instead of exporting agricultural products the country would export value added manufactured goods. Thus the overriding objective of India's trade policy had been of import substitution and self-sufficiency at least till the fifth five year plan. The domestic market was there for taking while export prospects looked bleak. Import competing industries were promoted behind high tariff walls and/or quantitative restrictions. Imports were regulated through a set of restrictions covering investments, foreign exchange transactions and international trade. This strategy achieved limited success but at a high cost due to the fact that - as domestic production expanded, there was a growing requirement of raw materials, components and capital goods which could not be met indigenously. The requirement of petroleum and fertilizers, indigeneous production of which was limited, was also increasing. This resulted in a significant pressure on the need for imports. And with exports growing at a much smaller rate than imports, the needed foreign exchange for imports was made available by the government at highly favourable terms to these industries. The result was imbalances in India's Balance of payments and heavy international debts.

From the middle of 1970s, it was recoganised that an import substitution policy by itself could not bring about a viable balance of payment situation and there was a need for vigorous export promotion policy. The import policy was gradually relaxed for import of raw materials and components for domestic production. Simplified procedures were introduced for importing technology, raw materials and capital goods needed for export production. Export processing zones were established and agencies set up to promote exports. An important change in the policy was introduced in 1978-79 when all items not specifically restricted or banned were listed under the open general license. It may, however, be mentioned that all these measures were to promote exports in the manufacturing sector. It was implicitly assumed that the country should export value added goods and manufactured goods. Trade in agricultural products was excluded from the trade policy regime whether we consider its rationale or structure.

7.2.2 Trade Policy during the Sixth Five Year Plan

Liberalization in trade policy has been pursued during the eighties to promote technological upgradation in the industrial sector. Periodic currency adjustments were carried out in exchange rate management to facilitate export growth. The basic goal of trade policy from the beginning of sixth five year plan was of stimulating export led growth by increasing the share of international trade in GDP. Self-sufficiency - though still an important policy objective - was redefined during this plan. The plan envisaged promotion of exports and invisible earnings in an effort to secure self-reliance. It was only during the sixth five year plan that the government policy, while emphasizing stable policy environment for exports and production in general, emphasized the importance of agricultural products also. A shift from an inward to a more outward-oriented policy and inclusion of agricultural commodities in trade policy regime in the sixth five year plan is clear from the following abstracts from sixth five year plan [1].

A major task facing the country is to reduce its dependence on imported energy and generally to promote exports and invisible earnings in an effort to secure self-reliance. It should be obvious, but often is not, that self-reliance does not mean self-sufficiency in all sectors of the economy. So long as the country is able to pay its way it cannot be said to be dependent on others. The choice of the outputs to be produced domestically and those to be imported should then depend broadly on economic calculus and the long-term comparative advantage which the economy enjoys. A Sustained increase in exports over a period of years cannot be achieved in the absence of a stable policy environment governing exports as well as production for export and production generally. Frequent changes in policies create uncertainty which is detrimental to the establishment of a stable market abroad and to risk taking inherent in investment decisions. The environment for production also has to be such as to enable enterprising individuals, agencies and corporations to exploit the available opportunities to the full. Except in very special cases, any conflict between objectives must therefore be resolved in favour of exports.

The success in the export effort of the country on the scale visualized in the plan will also depend to a great extent on augmenting agricultural production. All available instruments of policy will have to be geared to the promotion of the proposed increase in agricultural production. An important task of policy is to ensure that the gains of the technology and publicly supported programmes accrue increasingly to small and marginal farmers and are reflected in the adequacy of remuneration for agricultural labour. Agriculture is increasingly becoming science based, and future advances in application of science to agriculture will need to concentrate on improving productivity in agriculture consistent with the need to conserve non-renewable sources of energy.

The agricultural growth pattern during the sixth plan period has to take into account the immediate as well as the long-term needs of agricultural commodities both for domestic consumption and for exports. The highest priority will be accorded to bridging the gap prevailing between actual and potential farm yields even at current levels of technology through the removal of the constraints responsible for this gap. The untapped yield reservoir is quite high in most of the farming systems in the country and thereby serves as a source of optimum for achieving accelerated growth. Both agriculture and fisheries will have to receive concurrent attention through the development of appropriate packages of technology, services and public policies, which can help to enhance production from the soil as well as the sea and thereby improve the income of farmers and fishermen. For achieving greater efficiency of farm management, attention to non-monetary inputs is as essential as to cash inputs.

7.2.3 Trade Policy: Seventh Five Year Plan

The objective of a breakthrough in exports was considered to be a key element of the foreign trade and payments strategy during the seventh five year plan. However, a clear distinction was made between the agricultural sector and industrial sector. Trade in agriculture (like the trade policy regime of first five year plan to fifth five year plan and unlike the trade polciy regime of sixth five year plan) was excluded from trade policy regime of Seventh Five Year Plan. It was stated in the plan that, even in the long run, export growth will have to rely on manufactures and more important still on an increasingly wider spectrum of products. The plan assumed that unlike in the case of agricultural commodities, output growth of manufactures is not constrained by the availability of land, while shortage of raw materials and other inputs can be overcome through recourse to imports. It also presumed that market penetration becomes easier because of their exceedingly small share at present in the world trade in manufactures. A sizeable contribution of increase in exports (envisaged at 6.8 percent per annum in volume terms) was targeted to be attained by industrial exports during the plan. The objective of the agricultural sector was to achieve self-sufficiency in foodgrains and removal of poverty and unemployment in the rural sector. The plan stated, "The agricultural sector is expected to grow at an annual rate of 4.0 percent in gross output and 2.5 percent in terms of value added. This high growth rate is justified on the basis of likely demand generation and of maintaining self-sufficiency in foodgrains. The growth in agricultural production if properly directed, can reinforce the attack on poverty and unemployment. A breakthrough in exports leading to a substantial step up in real growth of export earnings is a key element of the foreign trade and payments strategy retained in the seventh plan. Industrial products will contribute sizeably to the attainment of the target set for exports. In fact, even in the longer run export growth will have to rely on manufactures and more important still, on an increasingly wider spectrum of products". The projected volume growth for another class of commodities which include, among others, tea, spices, cotton textiles, leather and leather manufactures and marine products will not,however, keep pace with the growth of GCP. Such exports are beginning to encounter supply constraints because of increasing domestic demand or inadequate output growth, or both. Export targets set for them are rather modest based as they are, on realistic assumptions about feasible expansion of production in medium term through measures to improve productivity, for the full effect of investment to strengthen the production base may be felt only after a period of time. Comparatively modest targets have also been retained in the case of jute manufactures, cotton textiles, iron ore and unmanufacturing tobacco, where the limits to export growth are set primarily by world demand"[2].

Foreign trade and payments strategy of a break-through in exports during the seventh plan period could not help in easing the balance of payment situation which had been continuously under strain for the past several years. The current account deficit increased to Rs. 8457 crores per annum during the seventh plan period as against an average of Rs.2616 crores during the sixth plan period. Though the robust growth in exports in the last three years of the seventh plan did lead to some improvement in balance of payment position in 1989-90, but it deteriorated again in 1990-91. The current account deficit during 1990-91 was estimated to be Rs.13942 crores. The worsening in the current account deficit in 1990-91 was partly on account of a trade deficit (due to Gulf War) which was marginally worse than the previous years but mainly on account of deterioration in the invisibles account because of adverse developments in the capital account. The political uncertainty at home, coupled with rising inflation and widening fiscal deficits led to erosion of international confidence in India. This resulted in higher interest payments, decline in commercial borrowings and outflow of NRI deposits. By June 1991, the BOP crisis had become overwhelmingly a crisis of confidence in the governments' ability to manage it. A default of Payment had become a serious possibility.

7.2.4 Economic Reforms: 1991-92

Given the large current account deficit and the consequent need to finance it through large scale external borrowing, it became essential to restore international confidence. This required the management of the BOP and the budget in a manner which ensured viability of BOP and strengthened the debt servicing ability of the economy. To this end, a series of economic reforms were introduced in 1991-92.

A strategy of economic reform introduced in 1991-92 was based on a combination of measures aimed at economic stabilization as well as structural reform.

The macro-economic measures undertaken in 1991 and 1992 to stabilize the economy included steps to reduce the fiscal deficit, tighter monetary policy and the introduction of a unified market determined exchange rate. The exchange rate of rupee was adjusted downward by about 18 percent in two steps on July 1 and 3, 1991 followed by the abolition of cash export subsidies designed to improve export incentives and make them more uniform. Fiscal deficit was reduced by about two percentage points from around 8.4 percent of GDP to 6.5 percent of GDP in 1991-92 to be followed by continued fiscal consolidation. Simultaneously the government negotiated access to external finance from multilateral financial institutions - the IMF, the World Bank and the Asian Development Bank (ADB) - as well as bilateral donors to support the BOP over the next 2 or 3 years. In May 1991, the government leased 20 tonnes of gold out of its stock to the State Bank of India to enable it to sell the gold abroad with an option to repurchase it at the end of six months. In addition, the government allowed the Reserve Bank of India to ship 47 tonnes of gold to the Bank of England in July. This helped to raise about $ 600 million.

The macro-economic measures were accompanied at an early stage by a paradigm shift in trade and industrial policies (since June 1991) with a view to integrating the Indian economy better with the rest of the world. The important features of the new trade policy - announced in July 1991 - include greater reliance on exchange rate management, elimination of product specific export incentives, reduction in tariffs and substantial elimination of licencing as well as other restrictions. One of the important measures undertaken as per this policy was that the value of the rupee was adjusted downward by about 20 percent in July 1991. This was followed by a liberalization of the foreign trade regime through some reduction in the quantitative restrictions. The import policy regime was revamped by shifting a significant number of items outside the purview of import licencing. Exporters were given entitlements equal to 30 to 40 percent of their export earnings in the form of EXIM scrips [3] against which even restricted items were allowed to be imported. Along with these, the government dispensed with a number of export incentives including the cash compensatory support to exports. As a first step towards a gradual reduction in the tariffs, the Union Budget for 1991-92 reduced the maximum rate of import duty from more than 300 percent to 150 percent.

Within eight months of introducing these trade policy changes, the government initiated another set of policy changes in the areas of trade and trans border capital flows coinciding with the Union Budget for 1992-93. The maximum tariff rate was further lowered to 110 percent. The EXIM scrip scheme was replaced by a system of partial convertibility of the rupee on the current account of the Balance of Payments. Under the new system, all foreign exchange remittances, whether earned through exports of goods and services or remittances, can be converted into rupees in the following manner: 40 percent of the foreign exchange remitted could be converted at the official exchange rate while the remaining 60 percent at market determined rate. The foreign exchange surrendered at the official exchange rate will be available to meet the foreign exchange requirements of essential imports such as petroleum and oil products, fertilizers, defence and life saving drugs. All other imports, except for a "negative" list are freely importable provided the foreign exchange for these imports were obtained from the market. Similarly, the foreign exchange required for other payments on private account including travel, debt service payments, dividends, royalties and other remittances were required to be obtained at the market rate. In another significant step, to arrest the diversion of foreign exchange to illegal channels, the government had legalised the import of gold and had introduced a Gold Bond Scheme, with a view to help mobilise the idle private gold reserves of the country to supplement official reserves. The important measures undertaken as per the new industrial policy - announced in July 1991 - has abolished industrial licencing for all but 18 product categories such as defence equipments, industrial explosives, electronic, aerospace, coal, petroleum, alcohol, hazardous chemicals, pharmaceuticals and drugs, and certain consumer goods such as sugar, edible oils, refrigerators, motor cars and consumer electronics. The maximum asset limit on the size of the companies which had hitherto been enforced under the Monopolies and Restrictive Trade practices Act, has been scrapped. As a result, firms will be able to grow to optimum size and reap the benefits of economies of scale. The government has also abolished the phased manufacturing programme under which domestic manufacturers were hitherto required to increase the domestic input-content of their products in a specified time period.Areas reserved for the public sector were narrowed down from 17 to 8 and greater participation by the private sector was permitted in core and basic industries. Along with a reform of industrial policies, steps were taken to facilitate the inflow of direct private investment. The limit of foreign equity holdings was raised from 40 to 51 percent. Technology imports for priority industries are automatically approved for royalty payments upto 5 percent of domestic sales and 8 percent of export sales or for a lumpsum payment of Rs. 1 crore. To enable the public sector to work efficiently, the public sector units have to be given the greatest autonomy in their operations. A system of full responsibility and complete accountability will have to be enforced on public sector managements.

The result of the above policy reforms and successful mobilization of exceptional financing was a marked improvement in the build up of foreign exchange reserves by an additional amount of US $ 3.57 billon during 1991-92 from US $ 1.1 billion in June 1991. The increase in reserves combined with stabilization and structural reforms restored international confidence and also provided the basis for further liberalization of trade, tariff, export credit and foreign investment policies during 1992-93. The year 1991-92 closed with important changes in trade and exchange rate policies.

7.2.5 Export-Import Policy (EXIM Policy) for the Eighth Plan Period (1992-93 to 1996-97)

The Export-Import policy contains, apart from the negative list of imports and exports, all measures in force for augmenting export production and exports, the prescribed procedures or conditions (where applicable) for exports and imports and the performance requirements for duty free or reduced duty imports for export production. The export-import policy (EXIM policy) for the Eighth plan period, announced subsequent to the Union Budget 1992-93, has operationalized many of the trade policy changes announced in the budget and has given a further push to liberalization..

7.2.5.1 Objective of New Trade Policy: The general trade policy objective is to make Indian industry competitive and efficient by exposing it to greater competition, and to integrate it with the global economy. Its principle objectives are: (a) to establish the framework for globalisation of India's foreign trade; (b) to promote the productivity, modernisation and competitiveness of Indian industry and thereby to enhance its export capabilities; (c) to encourage the attainment of high and internationally accepted standards of quality and thereby enhance the image of India's products abroad; (d) to augment India's exports by facilitating access to raw materials, intermediates, components, consumables and capital goods from the international market; (e) to promote efficient and internationally competitive import substitution and self-reliance under a deregulated framework for foreign trade; (f) to eliminate or minimise quantitative, licensing and other discretionary controls in the framework of India's foreign trade; (g) to foster the country's Research and Development (R&D) and technological capabilities; and (h) to simplify and streamline the procedures governing exports and imports.

7.2.5.2 Trade Policy Measures: The new EXIM policy has now specified the "negative" list of imports, the import of which is either banned, canalised or subject to import licence. Similarly, the EXIM policy has specified a "negative" list of exports, the exports of which is either banned, canalised or subject to export licence. In addition to specifying the negative lists of imports and exports, the EXIM policy has announced a set of incentives for exports and deemed exports.

The negative lists of imports and exports were pruned. The lists of prohibited items, restricted items and canalised items for both exports and imports were narrowed.

The maximum tariff rate was further reduced to 85 percent in 1993-94 from 110 percent in 1992-93, 150 percent in 1991-92 and 300 percent before 1991-92.

The procedure for import of capital goods (CG) has been simplified. Except for the negative list of imports, all items (capital goods, raw materials, components, spares etc.) are freely importable provided this import is fully covered by equity or is upto 25 percent of the value of plant and machinery, subject to a maximum of Rs.2 crores. The requirement of licence for import of capital goods under the scheme of direct foreign investment upto 51 percent foreign equity in high priority industries has been done away with. Actual user requirements for the import of capital goods, raw materials and components under OGL was removed.

The advance licencing system for exports was simplified (by various changes in the trade policy like simplifying and speeding up the process of issuing licences) to provide exporters with better access to imported inputs at duty free rates.

Several measures were adopted to strengthen the development of export houses and trading houses and instrument of promoting exports.They will now be allowed to install not only their own machinery but also machinery taken on lease. These units have now been allowed to take up new activities in agriculture, horticulture, aquaculture, poultry and animal husbandry. Setting up of trading houses with 51 percent of foreign equity has been permitted for the purpose of promoting exports. Such trading houses would be eligible for all benefits available to domestic export and trading houses.

A salient feature of the EXIM policy 1992-97 has been its fundamental departure from the erstwhile trade policy regime for agriculture [4]. Some policy decisions as per the comprehensive amendments to the EXIM policy announced on 31st March 1993 - have been taken to promote exports in the agriculture and allied sectors. It was therefore decided that units engaged in agriculture, aquaculture, animal husbandry, floriculture, pisciculture, poultry and sericulture can avail of the benefits of duty free imports under the EOU/EPZ scheme even if they export 50 percent of their production; they can sell the remaining 50 percent in the domestic market, against the limit of 25 percent, permitted for the non-agricultural sector.

The definition of capital goods under the policy has also been widened to cover capital goods used in agriculture and allied activities so that units engaged in this sector can avail of EPCG [5] scheme for importing their equipment at concessional duty.

In addition, certain inputs and materials required by the agricultural sector such as prawn, shrimp and poultry feed; edible wax for waxing fresh fruits and vegetables; grape guard paper; dipping oil for treatment of grapes; wheat gluten; fish meal in powdered form; and grand parent stock (poultry) have been removed from the negative list of imports - they are now freely importable.

India's trade and sectoral policies are set in the framework of indicative five year plans and subsidiary policy documents. The EXIM policy (the new trade policy) announced subsequent to the Union Budget 1992-93 is co-terminus with the eighth plan. The eighth plan envisages, synonymous to the new trade policy, that a policy regime with fewer barriers to trade, both tariff and non-tariff, and which provides equal incentives for exports as well as production for the domestic market enables the countries to achieve not only impressive export growth, but also rapid and sustainable economic growth. A key objective during the eighth plan, therefore, was to move our trade policy regime towards greater liberalization for improved resource efficiency and increased international competitiveness. The agricultural sector, which until now (except sixth five year plan) retained self-sufficiency as a goal, could on balance, also be a major beneficiary of more liberal trading measures. The strategy for agricultural development in the eighth plan,thus, is envisaged to be aimed at not only achieving self-sufficiency in food but also generating surpluses of specific agricultural commodities for export. The total growth in exports, envisaged to be 13.6 percent per annum during the 8th five year plan, is to be achieved (unlike the seventh five year plan) by 15 percent growth in exports of manufactured products and 9.4 percent growth in exports of agricultural products. Thus the major tasks during the eighth plan period, as is evident from the following abstracts from the eighth plan document are (a) to operationalise the trade policy changes announced in the budget, (b) to carry out further trade policy reforms for better integration of the Indian economy with the rest of the world, and (c) to promote exports in the agricultural sector for better performance of exports and better allocation of resources. The Draft Eighth Five Year Plan [6] stated, "A key task during the eighth plan is to move our trade policy regime towards greater openness and to reap the full benefits of international trade. The process of reforms initiated recently represents only a beginning. We have to carry it further if we want to reap the full benefits of these reforms. Therefore, sustaining the pace of economic reforms will be the major challenge during the eighth plan period. While both macro-economic stabilization and structural policies are important to ensure an economic environment which is conducive to efficient growth of the economy, a whole set of sector specific policies also has an important bearing on efficiency and growth". It further observed, "The eighth plan will aim at consolidating the gains from the base built over the years in agricultural production, sustaining the improvements in productivity and production to meet the increasing demands of the growing population; enlarging the incomes of farmers, and realising the country's potential by stepping up agricultural exports. As the country increases production beyond its own requirements, or where the country has comparative economic advantage in the light of technological gains, or to promote diversification, the marketing of agricultural produce within and outside the country assumes greater importance.

Recent changes in the industrial policy with deregulation and lesser direction by the state, have thrown open opportunities for a rapid phase of expansion of the agro-based industries, particularly the processing units. Induction of the latest technologies in processing and better packaging can help Indian processed food products compete in the international markets, besides catering to the growing market within the country. Special efforts would have to be mounted to give a fillip to the production of fruits, vegetables, milk and meat for being processed in units established including those int he rural areas. The changes in the trade policies have vastly improved the prospects for realising the full potential of the country with its varied agro-climatic conditions from tropical to temperate regions, in producing commodities for exports. Apart from maximising the production of the traditional export commodities, the new seed policy has created a climate for producing non traditional commodities like flowers and different types of vegetables for exports. Systematic efforts to overcome such constraints as may exist in the development of markets abroad will have to be launched and sustained in the coming years. During the Eight Plan period, further trade policy reforms are needed to be carried out so that the economy is better integrated with the rest of the world. The key objectives of these reforms should be two-fold; (i) a further pruning of the "negative" lists of imports and exports, and (ii) a gradual reduction in both the level and dispersion of tariff rates".

7.3 GATT Agreement on Agriculture (Uruguay Round)

7.3.1 Introduction

The General Agreement on Tariffs and Trade (GATT) is a multilateral treaty encompassing rules and disciplines for an orderly world trading system. One of the major activities of the GATT, since its inception in 1947, has been launching of a series of Rounds of negotiations, to begin with for tariff reductions and later reduction of tariff and non-tariff barriers. The first round of GATT was held in 1947 which produced a charter for the international Trade organisation (ITO).Since then seven more rounds have taken place. The recently concluded eighth round, commonly known as "Uruguay Round" was initiated in September 1986, at Punta deleste, Uruguay, concluded in Geneva on Dec. 15, 1993 and finally signed on April 15, 1994. The trade negotiations involved 115 countries covering more than 90 percent of world trade [7]. The major objective of this round was to halt and reverse protectionism and to remove distortions in trade". Other objectives include: (a) accelerated economic growth of the developing countries through trade liberalization; (b) achievement of substantial trade liberalization; (c) elimination of unilateral actions outside the legal framework of the GATT system; and (d) strengthening of the GATT itself. The Uruguay Round negotiations were most ambitious and complex one as along with the traditional GATT issues (tariff and non-tariff barriers etc.) new areas like Trade Related Intellectual Property Rights (TRIPs), Trade Related Investment Measures (TRIMs), Trade in Services (GATS), Agriculture and Textiles were also covered in these negotiations.

7.3.2 Agreement on Agriculture

Agriculture was assiduously kept outside the domain of GATT for the seven rounds preceding the Uruguay Round. The trade distorting agricultural policies were ignored by GATT for all these years. Both developed and developing economies have practiced policies which led to a highly distorted international trade in agriculture. While the developed countries protected their agriculture through rigorous controls and subsidy system, the developing countries did the reverse - promote industrialization at the cost of agriculture. The extent of protection had been high uniformly both across products and countries. The Uruguay Round Agreement on Agriculture is an attempt to correct these distortions. The agreement establishes a basis for initiation process of reform of trade in agriculture with the long-term objective of establishing a fair and market oriented agricultural trading system. The reform processes are to be initiated by negotiations of the commitment of farm support and production through establishment of a strengthened GATT Rules and disciplines.

The commitments incorporated in the agreement on agriculture cover three distinct areas, namely, domestic support, market access, and export subsidies.

7.3.2.1 Domestic Support Commitments: Almost all developed countries provide positive support to their cultivators ranging from as high as 72 percent in the case of Japan, 37 percent in the case of EC and 26 percent in the case of USA [8]. It is provided that all domestic support in favour of agricultural producers (with the exception of measures that have been exempted from reduction) shall be reduced by 20 percent in the case of a developed country, 13.3 percent in case of a developing country. The reduction commitment shall be achieved in 6 years in the case of a developed country and in 10 years in the case of a developing country. The reduction commitments are to be expressed and implemented in terms of total aggregate measurement of support (AMS), or in terms of equivalent commitments, where the calculation of AMS is not practicable. AMS comprises two elements, namely, product specific AMS and non-product specific AMS. For the purpose of calculating domestic support to farmers, both product specific AMS and non-product specific AMS shall be taken into account. In the case of developed countries the obligation to reduce domestic support arises if the total AMS given to farmers exceeds 5 percent of the value of the total agricultural production of country. For developing countries, this limit has been set at 10 percent. The base period for calculating reduction commitment is 1986-88 [9].

An important feature of the Agreement is that Government service programmes are exempted from reduction commitment and therefore, shall not be taken into account for calculating the total AMS. These include (a) general services, such as, research programmes, pest and disease control, training services, extension and advisory services, inspection services, extension and promotion services, infrastructural services, including capital expenditure for electricity, roads and other means of transport, market and port facilities, water supply facilities, dams and drainage schemes, and infrastructural works associated with environmental programmes; (b) public stock holding for food security purposes; (c) domestic food aid; (d) direct payments to producers subject to their meeting certain criteria; (e) decoupled income support; (f) Government financial participation in income insurance and income safety-net programmes; (g) payments for relief from natural disasters; (h) structural adjustment assistance provided through producer retirement programmes; (i) structural adjustment assistance provided through resources retirement programmes; (j structural adjustment assistance provided through investment aids; (k) payments under environmental programmes; and (l) payment under regional assistance programmes [10].

Apart from the above mentioned generally available exemptions, developing countries have been provided three additional exemptions,namely, (i) investment subsidies which are generally available to agriculture; (ii) agricultural input subsidies generally available to low income or resource-poor producers; (iii) domestic support to producers to encourage diversification from growing illicit narcotic crops [11].

7.3.2.2 Market Access: Under market access commitments, all member countries of GATT are required to (i) replace all types of non-tariff barriers with tariff barriers. (ii) Tariffs resulting from this tariffication process as well as other tariffs on agricultural products are to be reduced by an average of 36 percent in the case of developed countries and 24 percent in the case of developing countries. The reductions are to be undertaken over 6 years in the case of developed countries and 10 years in the case of a developing country. (News of the Uruguay round; 1994-p. 8-9) [12]. The least developed countries are not required to reduce their tariffs. The agreement contains special safeguard provisions for tarrified products. It says that if the volume of imports of a particular product in a particular year and the price of the imported product falls below the average of 1986-87 price, the country concerned may impose a specified duty temporarily for one year to safeguard its agriculture. Another stipulation in the case of market access is that where there is no significant imports, minimum access opportunities will have to be provided i.e. closed markets for agricultural imports in developed countries will be opened up. These countries are required to import at least 3 percent of their domestic consumption and gradually raise it to a minimum of 5 percent over 6 years. Minimum levels of access have, however, not to be granted by countries which do not have obligation to tarrify non-tariff restrictions (such as those maintained for BOP reasons). However exemptions have also been provided for staple crops for which there is no minimum import requirement.

The proposals in the Final Act, which modifies the Dunkel proposals, provide that once the New Agreement is implemented, countries have to provide access opportunities to imports of at least 4 percent of their total consumption as between 1986 and 1988, except for those primary products which are considered as staple in the tradiational diet of a developing country. (In India perhaps rice and wheat could qualify as such.). In the latter set of commodities the minimum access opportunities would have to be 1 percent of the corresponding domestic consumption to begin with. The access opportunities for the non-staple commodities would have to be increased annually by 0.8 percent between the first and the sixth year. For staple commodities, the minimum access opportunities have to be increased to 2 percent at the beginneing of the fifth year of the implementaion period and up to 4 percent of domestic consumption, at the beginning of the tenth year after the enforcement of the Agreement. (Mukherji, I.N.) [13].

7.3.2.3 Export Subsidy Commitments: The GATT agreement calls for reducing the value of direct export subsidies to a level of 36 percent below the 1986-90 base period level over 6 years implementation period and the quantity of subsidised exports by 21 percent over the same period. In the case of developing countries, direct export subsidies are to be reduced by 24 percent and the quantity of subsidised exports by 14 percent over 10 year implementation period [14]. Export subsidies included in reduction commitments are: (a) direct subsidies contingent on export performance, (b) export sale of non-commercial stocks held by Government at a price lower than the price charged to buyers in the domestic market, (c) payments on the export of an agricultural product that are financed by virtue of government action, (d) subsidies to reduce the costs of marketing exports of agricultural products including handling, upgrading and other processing costs, and costs of international transport and freight, (e) internal transport and freight charges on export shipments on terms more favourable than for domestic shipments, and (f) subsidies on agricultural products contingent on their incorporation in exported products. The reduction commitments have to be made in terms of budgetary outlays as well as quantities.

As regards export subsidies not listed above, it is provided that they shall not be applied in a manner which results in, or which threatens to lead to, circumvention of export subsidy commitments.

In the case of developing countries, the agreement provides that during the implementation period they shall not be required to undertake commitments in respect of the export subsidies listed in (d) and (e) above.

7.4. Implications of Trade Policy Reforms on India's Agricultural Exports

7.4.1 Implications of GATT on India's Agricultural Exports

India has not undertaken any commitments under the Uruguay Round Agreement on Agriculture.

7.4.1.1 Domestic Support: In the case of domestic support, almost all developed countries provide the support to their cultivators (ranging from as high as 76 percent in the case of Japan, 52 percent in the case of EC and 43 percent in the case of USA) [15]. India, however, imposes net tax on its cultivators due to the fact that - (i) India does not provide any product specific support other than market price support. During the reference period (1986-89), India had market support programmes for 22 products, out of which 19 are included in the country's list of commitments filed in GATT. According to the estimate made by the Commerce Ministry, the total product specific AMS was negative (Rs. 244.9 billion) during the base period. The total non-product specific AMS came to Rs. 45.8 billion. Taking both the product specific and non-product specific AMS into account, the total AMS in case of India was minus Rs. 198.6 billion, equivalent to about - 18 percent of the value of total agricultural output [16]. (ii) over the years the AMS has not only been negative but has been increasing overtime and stood at Rs. - 341.44 billion during 1992-93. This massive increase of 73.77 percent in total AMS seems to be the effect of depreciation [17].

Since India's total AMS is negative and that too by a huge magnitude, the question of our undertaking any reduction commitment does not arise. As such, India has not undertaken any commitment in its schedule filed in GATT.

It may be mentioned that in the calculation of our domestic support, no account has been taken of the exemption of input subsidies to low income and resource poor farmers in developing countries. In India, operational holdings of 10 hectares or less account for 79.5 percent of agricultural land. If farmers holding less than 10 hectares of land are considered to be resource-poor or low income, almost 80 percent of the input subsidies will qualify for exemption from inclusion in non-product specific AMS. This would result in a AMS figure being lower even [18].

7.4.1.2 Market Access: India, classified as a developing country with balance of payments problems, does not have any market access commitments for agriculture and is therefore free to maintain quantitative restrictions on imports. However, according to the Ministry of Commerce, countries with negative balance of payments would have to indicate the ceilling tariffs which would apply when BOP constraints are no longer there and quantitative import restrictions are consequently withdrawn. As such, india has indicated that the ceiling tariff levels would be 100 percent for primary products, 150 percent for processed products and 300 percent for edible oils [19].

7.4.1.3 Export Subsidy: The GATT agreement on agriculture, as mentioned earlier, lists six types of subsidies to which the reduction commitments will apply. Indian exporters of agricultural products do not get direct subsidy. The only subsidies available to them are in the form of (a) exemption of export profits from the income tax which is not one of the listed subsidies in GATT agreement and (b) subsidies on costs of freight on export shipments of certain products like fruits, vegetables and agricultural products. This is part of the two listed subsidies for which developing countries are exempted from the reduction commitments. The total aggregate value of subsidies given to farmers namely, subsidies on fertilizer, electricity, seeds, persticides and cost of credit is well below the ceiling prescribed in the GATT Final Act. Thus, the GATT commitments framed in the subsidies code in case of export subsidies will not have adverse impact on agricultural exports from India. Thus, India may continue with the present subsidy and has got scope to even enhance it, if required, Further, the reduction of subsidy by developed countries may lead to improvement in world prices of agricultural products which may have a stimulating effect on agricultural exports of developing countries like India.

7.5 Conclusion

The analysis of GATT commitments in agriculture and their impact on India's agricultural exports reveal that the GATT agreement on agriculture does not restrain us from following our development policy with regard to agriculture as India has not undertaken any commitments under the Urgugay Round agreement on agriculture. Thus, to take full advantage of such opportunities of world trade in agriculture, the required strategy of agricultural exports as suggested in Chapter 6, should be to enhance policy initiatives in the areas of yield improvement and resource management for increasing export surpluses of this sector with least effects on domestic prices.

The analysis of Section 7.3 reveal that the recent trade policy reform has reduced tariff and non-tariff protection for the industrial sector, and sought to eliminate the over valued exchange rate through a substantial depreciation. This has led to removal of an implicit bias against agriculture which had led to a misallocation of resources for the past four decades and at present, has created a climate for encouraging agricultural exports in the country. However, the real effect of these policy initiatives on agricultural exports would depend on how far these policy initiatives are successful in not only creating a climate for increasing agricultural exports, but also providing consistency and stability to these exports by moving away from the stage of incidental activity (or peripheral and residual activity) to a planned deliberate effort as suggested in the strategy of agricultural exports (Chapter 6).

Notes

1. Sixth Five Year Plan : Govt. of india.

2. Seventh five Year Plan : Govt. of India.

3. EXIM scrips were freely tradeable and could be used to import any item in the limited permissible list, the non-sensitive canalised list, OGL items for actual users and non-OGL capital goods which were not in the restricted list.

4. The trade policy regime in India, over the past four decades, has made a clear distinction between agricultural sector and the industrial sector whether we consider its rationale or its structure (Deepak Nayyar, FAO Seminar, 1994).

5. EPCG - Export Promotion Capital Goods Scheme.

6. Eight Five Year Plan : Govt. of India.

7. GATT (1994) : Country Studies : India.

8. Economic Journal vol 103. p.427, p. 1515.

9. Final Act Embodying the results of the Uruguay Round of Multilateral Trade Negotiations, Marrakesh, 15 April , 1994.

10. Mohanty P.K. (1994) : Uruguay Round Agreement an Agricultre - An overview, Ministry of Commerce Govt. of India.

11. Ibid.

12. News of the Uruguay Round, 1994 p.8-9.

13. Mukherji, I.N. : The New World Trade Order : Access to Trade and Markets; Paper presented in the Seminar on The New World Trade Order : Prospects and concerns. Madras University, Dec. 1994.

14. News of the Uruguay Round : 1994 : p.10-11.

15. Economic Journal, vol.103, p. 427, 1515.

16. Mohanty, P.K., 1994, Uruguay Round Agreement on Agriculture - An Overview, Ministry of Commerce, Govt. of India.

17. Gulati Ashok and Sharma Anil, 1994 : Agriculture under GATT, Seminar on Uruguay Round Agreement on Agriculture IIFT, 1994.

18. Mohanty P.K. (1994) : Uruguay Round Agreement on Agriculture, Ministry of Commerce, Govt. of India, 1994.

19. GATT (1994) : Country Studies : India

Chapter 8

Conclusions

8.1 Introduction

Information concerning the pattern of international competitiveness and its tendency to change overtime is highly desirable input to decision making. Enhancing competitiveness may be termed as one of the major objectives of trade strategy. Competitiveness of exports may depend on whether the composition of exports, as suggested by trade theories, is determined by comparative advantage or not. The comparative (cost) advantage has two useful applications - (i) it serves as one of the fundamental concepts in descriptive (or positive) theories by providing a basic explanation of the international pattern of specialisation in production and trade (ii) it plays an important role in prescriptive (or normative) economics by providing guidelines for government policies on resource allocation and trade. In the positive role, trade theories are able to tell us why, after all the composition of exports differ from one country to another. The pattern of trade between different countries is determined by the pattern of comparative advantage among them.

The strategy of industrialisation that takes comparative advantage into consideration provides a suitable basis for an international division of labour between countries of different levels of economic development. An international division of labour on this basis is considered to be desirable objective in the long run interest of harmonious development of the international trade and the world economy. The neglect of comparative advantage very often leads to diseconomies of scale and inefficiency in production. Thus, comparative advantage consideration should be brought to bear upon the choice that has to be made regarding the kind of specialisation in exports.

However much of the research in this area has focused on trade in industrial goods and a number of studies have worked out the comparative advantage in manufacturing. In the case of agriculture, though the thrust has been shifting to agricultural exports since recently and a number of studies on performance, problems and prospects of agricultural exports are there, the study of comparative advantage in agricultural exports has received little attention. The analysis of comparative advantage in exports of agricultural products has received scant attention on the ground that agricultural products are intensive in the use of natural resources and thus the pattern of trade in such products depends to a considerable extent on the not easily quantifiable resource endowment of the country. The neglect, however, has also been partly to the fact that most of the studies on comparative advantage were done for developed economies (mainly USA and Japan) which were at that stage of development where primary exports weightage was quite low. As comparative advantage for developing economies seems to lie in agricultural products and labour intensive manufactures primary exports weightage seems to be quite high for low income predominantly agricultural economies of South Asia which necessitates the analysis of comparative advantage in agricultural exports of South Asia.

8.2 Hypothesis

South Asian economies, at present, confront a difficult situation in their international competitive position. The staggering trade deficits of the past several years and the borrowing needed to finance them obviously can not be sustained for long. Inspite of the efforts to diversify exports (along regional and commodity dimensions) and the introduction of numerous export promotion measures since late 1970s, the exports of South Asian countries failed to register growth rates either comparable to those of other developing nations or commensurate with the amount of effort and initiative taken by these countries. Export promotion as one of their strategies did not help them in adjusting to the crisis through increased export earnings. We hardly need to reiterate that their market share in world exports declined. They were outstripped by other developing countries who emerged more successful in gaining export market shares, in addition to what they already possessed.

The analysis of export performance of South Asian Countries as compared to NIC-4 and ASEAN-4 (Chapter 2) reveals that though most of the countries in South Asia shifted to relative export orientation policies since 1970s and have initiated structural economic reforms since mid of 1980's, lack of international competitiveness still seems to be the root cause of poor performance of exports in these economies. Thus, the process of trade liberalisation which has been gradual, slow and detrimental to growth due to lack of competitiveness may be made easier, quicker and more successful by adopting trade policies which are aimed at improving the underlying competitiveness of exports of South Asian countries. What is needed to make the position less costly is the development of a national competitiveness policy and the major objective of this trade strategy may be termed as enhancing competitiveness of exports.

The question arises, "What are the major corrections through which South Asian countries could improve their competitiveness of exports" ? In other words, we have to look at the determinants of competitiveness. An attempt is made to answer this question in Chapter 3.

The analysis of Chapter 3 reveals that competitiveness is a dynamic phenomenon. Though it depends on initial conditions, factors and resource endowment, size, location, social mores and international environment, it is to be evolved as an integral component of the development process. The reasons may be : First, competitiveness of exports may depend on whether the composition of exports, as suggested by trade theories, is determined by comparative advantage or not. Second, comparative advantage itself is a dynamic phenomenon as there might be shifts in the comparative advantage with the process of economic growth.

The sources of comparative advantage lay in factor endowments. Changes in factor endowments over a period of time (due to change in technology, scale economies etc.) will lead to change in comparative advantage. Thus, a priori one would expect underdeveloped economies having scarcity of capital and skill to possess a traditional production structure and thus specialise in production and export of raw materials and agricultural products. As economies move from lower to higher levels of development, they accumulate more capital, develop skills and generate better technology and thus move up the ladder of comparative advantage from specialisation in primary products to products which are capital, skill, technology and knowledge intensive.

In accordance with the principle of comparative advantage, Japan since 1950s, NIC-4 since 1960s and ASEAN-4 in 1970s expanded their exports and employment by specialising in highly productive agriculture in the initial stages of development, then in labour intensive products and then moved into capital intensive and skill intensive exports in later stages at development. Thus, we see that countries whose trade strategy supported pattern of trade that not only conformed to comparative advantage as it existed initially but also changed the strategy in accordance with shifts in the comparative advantage overtime improved their competitiveness in the world markets. NIC-4 (except Hongkong) having realised that an expanding industrial sector cannot be fuelled for long by resource transfers from a static agricultural sector and the economic potential of agriculture must be mobilised before it can be used, improved their competitiveness by increasing their share of primary exports in the initial stages of development. The success of ASEAN-4 in the 1970s in international trade and growth was also based on the same policy. By expanding agricultural exports

ASEAN-4 countries realised greater foreign exchange earnings and improved the linkages between agriculture and other sectors.

The mobilisation of agriculture for general development and trade would depend, as per the experience of successful Asian countries, on macroeconomic and trade policies which are fairly conducive to expansion of agricultural exports. Export growth has been particularly rapid in cases where more or less equal incentives have been provided to exports of different sectors and individual export products. Macro economic and trade policies in Korea, Taiwan and ASEAN-4 were fairly conducive to expansion of agricultural exports.

It may, however, not be implied that only protection will lead to better performance of exports by the agricultural sector of the economy. Unless a sector (industry or agriculture) receiving protection is pushed rapidly to become internally competitive by not only producing more but of quality too, it may not contribute significantly to improvement in competitiveness. The effects of primary exports on industrial development depend to a considerable extent on output relationships. In other words agriculture could be exploited for general development and trade only if it first produced a surplus which was there for exploitation.

Comparative analysis of growth of agricultural production by countries reveals that agricultural exports were growing fastest in the countries with the highest rates of growth in agricultural production.

Production increase will prove to be effective i.e. cost disadvantage will be reduced only if backed by adequate infrastructure. Infrastructure in the form of ports, railways and roads often represents important inputs for primary exports and their availability may contribute to the development of industrial activities.

The Government is expected to develop the infrastructure by providing various types of economic services for promoting overall developmental climate in the country.

Comparative study of NIC-4, ASEAN-4 and South Asian countries (Chapter 6) reveals that the share of economic services in total expenditure of Central Government (consolidated accounts) works out much higher for NIC-4, ASEAN-4 and China as compared to South Asian countries.

Thus it was hypothesised in Chapter 3 that countries whose trade strategy supported composition of trade that not only confirmed to the comparative advantage as it existed initially but also changed the strategy in accordance with shifts in comparative advantage (overtime) with major emphasis not only on subsidies but on production and infrastructure improved their competitiveness in world trade.

In the light of the above hypothesis we may state that the first requisite for enhancing competitiveness of exports is a trade strategy which supports a composition of trade that conforms to comparative advantage (static as well as dynamic). In other words, the first requisite for evolving competitiveness of exports is to identify the export sector and commodities which will give comparative edge in terms of resource endowment and capabilities of the country concerned. Having identified the sector and commodities which could easily be developed for trade and thus for which export effort would be worth undertaking the second requisite is the formulation of a trade strategy which is conducive to expansion of exports in the identified sector.

8.3 Basic Thrust of the Study

The basic thrust of the present study is that the comparative advantage for developing South Asian economies characterised by abundance of labour, scarcity of capital and heavy dependence on natural endowments would seem to lie in agricultural products and labour intensive manufacturers. Thus growth model based on primary and agricultural products seems to have considerable advantages in maximising agriculture's contribution to overall development. Further, South Asian economies are basically agricultural in nature. Despite the rapid growth of industrial production, their dependence on agriculture is staggering. At present, about 50% of national income of Sri Lanka, 40% of India and 25% of Pakistan is derived from agricultural sector. There are several reasons why agricultural exports should contribute a similar proportion to total exports:

(a) domestic Resource Cost of each rupee of foreign exchange earned is lower in agriculture than in industry.

(b) factor incomes generated at home per rupee of output exported, directly and through backward linkages is higher for agriculture compared with manufacturing and mining.

(c) basically, agricultural commodity exports require the least amount of cash assistance. Foreign exchange earning is net in most of the agricultural commodity exports.

(d) world demand for some of the higher value agricultural products such as dairy items, fruits and vegetables and marine products has been increasing since last few years.

(e) since recently the thrust has been shifting to agricultural exports and changing international environment is also favourable to this.

(f) South Asian countries enjoy enormous opportunities in the developing region especially in Asia and Africa because of similarity in socio-economic conditions, initial stages of their economic development and the geographical proximity to the region.

(g) South Asian countries enjoy a fairly large potential for agricultural, horticultural and marine products. Despite these advantages, the share of these countries in global imports of agricultural products pales into insignificance (less than 1% of world agricultural exports). To look into the explanation for this paradoxical situation an analysis of comparative advantage of exports in agricultural products of South Asia has been attempted in this study.

The fundamental premise upon which this study is based is that comparative advantage considerations should be brought to bear upon the choice that has to be made regarding the kind of specialisation in exports of developing economies of South Asia. Thus the objectives of the study has been :

(1) To prove empirically that

(i) The comparative advantage for South Asian countries lies in exports of agricultural products and labour intensive products.

(ii) The sectoral composition of South Asian countries trade however has not been based on comparative advantage considerations, and

(iii) To stress the need (on the basis of empirical approximation of (i) and (ii) to increase share of agricultural exports in the export basket of South Asian countries to enhance competitiveness of their exports.

(2) To analyse the present position and problems of agricultural exports.

(3) To identify agricultural commodities (at the 3 digit SITC level) which will give South Asian Economies a comparative edge in terms of their resource endowment and capabilities and thus for which export effort would be worth undertaking.

(4) To offer Suggestions as to how agriculture can be exploited in a better way for trade in these economies. i.e. suggest a suitable strategy of agricultural exports.

The analysis proceeds in three stages : Chapters 1, 2 and 3 provide the theoretical background. Chapter 4 and 5 are empirical in content and deal with the empirical verification of pattern of trade of South Asian countries. Chapter 6 attempts an empirical analysis of agricultural exports at micro level by taking a case study of India. It analysis the performance of agricultural exports in India. Identifies the agricultural commodities (at three digit SITC level) having comparative advantage in India, analysis their export performance in comparison to their revealed comparative advantage and attempts to suggest a suitable strategy for increasing the share of agricultural exports (specially in the identified commodity groups) in the export basket of India to help enhance competitiveness of exports in India.

8.4 Findings of the Study: Chapters 1 to 7

Introduction serving as Chapter 1 outlines the approach for analytical purposes, reviews the literature and specifies the objective, scope and methodology of the study. The analysis of export performance of South Asian countries (Chapter 2) revealed that the root cause of poor performance of exports in South Asian countries has been lack of international competitiveness. Thus enhancing competitiveness may be termed as one of the major objectives of trade strategy for South Asia.

Chapter 3 is theoretical in nature and deals with the model of export determinant. It is divided into three parts : (a) Determinants of competitiveness (b) Sources of comparative advantage i.e. theories of comparative advantage and empirical models review and (c) theoretical validity of the H-O model of trade.

Looking into the determinants of competitiveness i.e. the major corrections through which competitiveness could be improved it is hypothesised in Chapter 3 that countries whose trade strategy supported composition of trade that not only confirmed to comparative advantage as it existed initially but also changed this composition in accordance with shifts in comparative advantage (overtime) with major emphasis not only on subsidies but on productivity and infrastructure.

The competitiveness of exports as per the first part of the hypothesis is to be evolved with development process by changing the composition of trade in accordance with changing or dynamic advantage. For this we have to provide an understanding of the historical forces which gives rise to current pattern of comparative advantage and provide an insight into likely direction of change in the future i.e., we have to look into the sources of comparative advantage. Thus before proceeding to the empirical analysis of comparative advantage in exports of South Asian countries, a review of theories of comparative advantage and empirical models review has been attempted in chapter 3 only to determine the model of export determinant.

A review of the theories of comparative advantage and the empirical models review (Chapter 3) reveals that the H-O Model has conventionally been used to explain the pattern of international trade. According to this theory (assuming the technology to be the same across countries), comparative advantage is the result of the resource endowment structure of different economies. A series of interesting hypothesis and modifications of the simple H-O theorem examining the importance of various alternate determinants of pattern of trade have been developed over the last forty years. A detailed review of these modifications reveal the following : First, sources of comparative advantage lay in differences in factor endowment. The weight carried by different factors may, however, vary from country to country and from sector to sector. Differences in natural endowment (land, labour and natural resources) have been taken to be as sources of comparative advantage in case of raw-materials and agricultural products. Capital (physical and human), technology, scale economies, research and development have similarly been given high weight in manufactured products. Second, in a world of rapidly changing technology one needs to focus on the dynamics of comparative advantage as compared to the static one. Third, the role of scale economies in determining pattern of trade would depend on the nature of trade (inter-industry or intra-industry) which in turn is dependent upon country similarities. "Countries with similar factor endowment will engage in intra-industry trade, while countries with different endowment will engage in H-O trade (Krugman, 1990). As there as substantial differences in the endowment pattern of developing economies of South Asia, a strong H-O component should be present among the determinants of the commodity composition of trade in these economies. The comparative advantage of trade in agricultural products of South Asian countries, thus, is examined (in static as well as in dynamic sense by practical application of the commodity version of H-O model.

The familiar neo-classical version of factor proportions explained the pattern of trade for the economy as a whole i.e. considered all the sectors - agriculture, industry etc. The later theories, except a few (Leamer, Linder) considered sources of comparative advantage in manufacturing sector only. The reason may be that most of the studies were for developed economies (mainly USA) who were at that stage of development where primary exports weightage becomes quite low. South Asian economies being underdeveloped economies characterised by abundance of labour, scarcity of capital, technology etc. our analysis will consider not only the manufacturing sector but all the sectors in these economies to judge whether their pattern of trade has been as per the resource endowment structure as reflected in the comparative cost advantages.

An attempt is made in Chapter 4 and 5 to verify empirically whether the pattern of trade of South Asian economies conforms to the structure of comparative advantages as reflected in their resource endowment structure (in static as well as in dynamic sense). Chapter 4 attempts to examine the proposition that the comparative advantage for South Asian countries lies in the exports of agricultural products and labour intensive manufacturers. The proposition is examined, as per the theoretical analysis of Chapter 3, by the empirical verification of H-O model of trade. Having established that the comparative advantage for South Asian countries lies in exports of agricultural products and labour intensive manufactures, we analyse, in Chapter 5, the composition of exports of these countries vis-à-vis comparative advantage to see whether the composition of exports of these countries has been as per comparative advantage or not.

The analysis of Chapter 4 takes in account both the static and dynamic nature of comparative advantage. The static nature of export compatibility is analysed by an application of the commodity version proposition of the Heckscher-Ohlin (H-O) theory. The dynamic comparative advantage is observed by testing empirically the factor proportions model over different points of time in an economy that is observing the model as a static one at different dynamic points. As the concept of comparative advantage involves intercountry comparisons, we attempt a comparison between South Asian countries, and some of the successful developing Asian countries, NIC-4 and ASEAN-4.

The Methodology adopted is as follows :

According to H-O model, comparative advantage (assuming the technology to be same across countries) is the result of resource endowment structure of different economies. Empirical application of the HO model requires a three way relationship between factor abundance, factor intensity and trade composition of the country concerned. The method adopted in the present study is based on the studies by Hirsch (1974) and Leamer (1984).

Comparative advantage according to Hirsch (1974) is the outcome of interaction between two factors, namely, factor endowment, which is a country characteristic, and factor intensity which is an industry or product characteristic. Comparative advantage is empirically taken to be high when the values of the factor intensity and the factor endowment go together. Comparative advantage is taken to be low in other cases.

We first estimate factor endowment of South Asian countries and also establish factor intensity for three sectors namely, agricultural, labour intensive manufacturers and capital intensive manufacturers. The interaction between these two factors i.e. factor endowment and factor intensity enables us to determine comparative advantage for South Asian countries in static sense. The dynamic comparative advantage is examined by considering the period between 1970 and 1985 and estimating the relative changes in factor endowment of South Asian countries as compared to NIC-4 and ASEAN-4.

A rigorous analysis of the patterns of comparative advantage however requires cost comparisons across countries. But the data required for such an analysis are rarely available. The export performance index which reflects all types of costs and non-price factors and is based on the empirical data pertaining to trade flows has often been used as an indicator of comparative advantage. Since this is revealed by the observed pattern of trade flows, it is called revealed comparative advantage (RCA). We work out the RCA index for South Asian countries, NIC-4 and ASEAN-4 to examine the performance of agricultural exports of South Asian countries and to identify agricultural products having comparative advantage in India.

As per the empirical approximations of Chapter 4 the South Asian countries are found to be relatively abundant in land and unskilled labour and thus the comparative advantage for South Asian countries as per the factor endowment proposition of comparative advantage (in static as well as in dynamic sense) lies in agricultural and allied products and labour intensive manufactures.

The analysis of Chapter 5 reveals inconsistency between composition of trade and comparative advantage. The comparative advantage for South Asian Countries, as per neo-classical doctrine, lies in agricultural products and labour intensive products. The composition of exports of these countries (specially India Pakistan and Bangladesh), however depict high percentage of capital intensive exports. Though as regards labour intensive exports there was some awareness on the part of these countries and the share of labour intensive exports showed little upward trend from beginning of 1980s, the share of agricultural exports showed decline not only in 1970s but even in 1980s.

In an attempt to analyse the above inconsistency between composition of exports and revealed comparative advantage in South Asian countries, empirical analysis of agricultural exports is attempted in Chapter 6 at micro level by taking a case study of India.

Chapter 6 is divided into three parts. The importance of agricultural export strategy for India is attempted in part I. The analysis of performance and problems of agricultural exports in

India is attempted in Part II and Part III suggests a suitable strategy of agricultural exports in India. The strategy of agricultural exports in India in the light of the hypothesis is further subdivided into two parts.

(a) The export planning and policies require as a first step, identification of export sector and the commodities which would give us a comparative edge in terms of our resource endowments and capabilities. Thus identification of agricultural products having comparative advantage in India is attempted in part (a).

(b) Having identified the agricultural products having comparative advantage in India we proceed with the second requisite of enhancing competitiveness of exports as per hypothesis of chapter 3, i.e. how to create export surpluses in the agriculture sector in part b. In other words the suggestions for better performance of agricultural exports in India is attempted in part b. The analysis of chapter 6 revealed that at present, India confronts a difficult situation in its international competitive position. The staggering trade deficits of the past several years and the borrowing needed to finance them obviously can not be sustained for long. The long term solution for India, thus, is to strengthen its export base in order to gain increased export earnings. The feasibility of continuing to rely on manufacturing export led growth as a major development dynamic is to be reassessed at this critical juncture of India's economic development. The export potential of the agricultural sector needs to be rediscovered to augment exports in India. The relative decline of agricultural export as compared to manufacturing sector during the 1970s and 1980s can be revived in mid nineties and 2000 as (a) India still has comparative advantage in exports of agricultural products, (b) the domestic resource cost of each rupee of foreign exchange earned is lower in agriculture than in manufacture, (c) factor incomes generated at home per rupee of output exported (directly and through backward linkages) is higher for agriculture as compared to manufacturing and mining. Net foreign exchange earnings is positive in most of the agricultural commodity exports, (d) world demand for some of the higher value agricultural products such as dairy items, fruits and vegetables and marine products has been increasing for the last few years (e) India enjoys a fairly large potential for agricultural, horticultural and marine products due to large geographical area with varied climates, temperatures and variety of soils accompanied by the largest coast lines in the world.

The performance of India's agricultural exports has, however, been quite poor (a) in comparison with other countries in the study (b) in industrial exports and (c) in world agricultural exports. The share of agricultural exports in India's total exports showed decline from 40 percent in mid 1960s to 18 percent by mid of 1990s. A slight improvement in this share was visible in 1995-96 (19.2 percent). Similarly, the share of India's agricultural exports in world agricultural exports declined from 2.0 percent in mid 1960s to less than 1 percent by end of 1980s and early 1990s. There was marginal improvement in this share to little above 1 percent in 1995. The explanation for this paradoxical situation lies in the macro economic policies on the one hand and lack of export surpluses on the other. The analysis of trade policy reforms in India (Chapter 7) reveals that though the implicit bias against agriculture has been removed and a climate for increasing agricultural exports has been created by reduction in tariff and non-tariff protection for industrial sector and eliminating the over valued exchange rate through a substantial depreciation. However, the real effect of these policy initiatives on agricultural exports and their contribution to general trade and development in the economy would depend on how far these policy initiative are successful in not only creating a climate for increasing agricultural exports but also on (a) whether the increase in agricultural exports are based on comparative advantage considerations and (b) supply management in agricultural sector i.e. to increase agricultural production significantly to meet not only the domestic requirements in full, but also to generate adequate export surpluses.

8.5 Conclusions and Suggestions

Focus on Selected Crops: The declining share of India's exports of agricultural commodities (having comparative advantage) in world market (Table 6.10) may indicate that these commodities received less attention in realising their export potential. The export promotion policies, especially subsidies have been directed towards sectors which did not possess comparative advantage. The import substituting crops benefited not only from increasing yields due to improved technology but also of price benefits (due to Government's support measures). The result was that not only did yields increase less for export crops but also that there was some shift of area from the export crops to the import substituting ones e.g., area for production of soybean increased from 2667 HA in 1989-91 to 6350 HA in 1998. The area under cereal crops decreased from 102278 HA to 99450 HA and rice from 2 to 1 for the same period. Thus, we may state that apart from maximising the exports of traditional commodities and maintaining their share in world exports, importance of promoting non-traditional exports with high growth prospects in world markets and high production potential at home such as horticulture, apiculture, fisheries and sericulture can hardly be overemphasised. The exports of non-traditional commodities like fruits, vegetables, flowers and honey require greater attention in realising the country's potential of agricultural exports.

Improve Crop Yields: The effects of primary exports on industrial development depend to a considerable extent on output relationships. The main source of export growth in developing countries was the rise in domestic output. This implies that a significant increase in agricultural production (specially of the products having comparative advantage) is the foremost prerequisite in exploiting India's agricultural export potential. The basic push has to come through (a) yield productivity improvement and (b) efficiency in resource management.

India's production potential (through yield improvement) in tea, spices and specially fruits and vegetables is enormous and it will be quite easy to generate a sizeable export surplus. For example, a large number of horticultural crops suffer from low yields (productivity) as compared to Asian countries' average and world average. If this is corrected per hectare yields would increase by 100 to 200 percent to take care not only of the increase in domestic consumption but at the same time export surpluses will be generated (Table 6.11). Prerequisites for productivity improvement, development of human capital and basic infrastructure and availability of producer (economic) services are the outcomes of public investment. Increase in public investment, thus, will lead to increase in productivity in the agricultural sector. In the Eighth Plan investment in most of the infrastructure sectors has fallen well below the target. It is doubtful that with a considerably lower level of public investment (33 percent) in the Ninth Plan compared to 43 percent (targeted) and 36.5 percent (anticipated) Government of India would be able to provide adequate infrastructure and producer services.

According to a Report of the working group on Agro-Food Processing Industry (Eighth Five Year Plan), nearly 20 to 25 percent of fruits and vegetables (excluding substantial losses in handling & transport) perish each year in India. Thus if 30 percent more is added to the yield improvement (which may be realised by eliminating the present wastes by proper planning at the production and post harvest phases and organisation of trade on a systematic basis) the increase in production may come out to be still higher.

Efficient use of Productive Resources: In most countries, the major challenge is how to intensify agriculture sustainably since appropriate land resources do not exist for horizontal expansion. Intensification could be achieved not by using them more efficiently. Improved efficiency should reduce waste and loss.

The problem of waste is universal. While unused land and water resources are often looked upon as "reserves", in reality their idleness represents a loss of the services which they could render to humanity if managed well. Thus, for the war on waste to be waged successfully, the mobilisation of all resources must be carefully phased and integrated.

A comparative study of agricultural growth in China, Korea and India shows that China and Korea could achieve higher rates of growth in agricultural productivity and yield by adapting strategy of full use of resources by waging war on waste. Whereas, in case of India inefficiency in the use of resources seems to be a major cause of low production in agriculture.

Every silver of land in China is cropped. New arable land is being created through terracing and levelling of hills. Even roof top plots are used. What cannot be used for agriculture is reforested. Chinese also appear to have developed the art of utilising rivers and irrigation canals for raising food, organic matter, feed soil amelioration, fish ducks and geese. The policy of full use of resources along with land extends to the conservation of water and recycling of organic wastes. Water conservation is the lifeline of agriculture in China. Also, the soil is carefully nurtured. Yearly enormous amounts of compost and organic material is mixed with soil. Availability of chemical fertilisers is still limited and hence use of organic manure. Organic recycling is obviously a key factor in yield increase.

In the case of Korea, an urgent task for the upliftment of people's standard of living and for meeting the increased demands of the state for agricultural products, has been to develop measures for the effective use of the country's land and water resources. A huge area of tidal land have been reclaimed by energetically pushing ahead with reclamation projects. Also, a mass campaign has been launched to acquire new land through land rezoning and the reclamation of hills. In this way, the country's soil productivity is increasing and great achievements have been attained in increasing agricultural production. The question of exploitation and utilisation of water resources is particularly important in agricultural production. The Korean policy is that only when irrigation systems are properly established through active development of water resources is it possible to make effective use of land resources for stable farming and ensuring high yields.

A major cause of low productivity in agriculture in India is the inefficiency in the use of resources-land and capital. Improvements in capacity utilisation and efficient project implementation in all areas, especially in irrigation, power, transport and food processing units are essential for achieving increases in productivity of capital in the agricultural sector.

Lack of post harvest infrastructure account for 20-40 percent losses at different stages of storage, grading, packing etc. (9th Plan, P. 57). Thus, if 30 percent more is added to the yield improvement (which may be realised by eliminating the present wastes by proper planning at the production and post harvest phases and organisation of trade on a systematic basis) the increase in supply may come out to be still higher.

Efficiency in the use of capital has often been low in the past because of delays in implementation and insufficient attention paid to efficient management. In 1993 there were 4132 units in the fruit and vegetable processing industry with an installed capacity of 12.60 lakh tonnes. But the capacity utilisation was as low as 44.4 percent with production of 5.59 lakh tonnes. Fruit and vegetable processing industry at present is utilising only about 13 percent of total production for processing as against the performance of other countries- Malaysia (83 percent), Philippines (78 percent), USA and Brazil (70 percent).

Efficiency in the use of land has been low due to the fact that more than half of the estimated land area of 328.8 million hectares remains utilised in India. It is necessary to adopt commercial approach for the wasteland development programme to make best possible use of wasteland resources.

Utilisation of wastelands for horticulture along with afforestation seems to be efficient use of land resources as (a) The wasteland which is unsuitable for field crops can profitably be utilised for horticultural crops, (b) horticulture pursuits are labour intensive, provide high income per unit of land and help in maintaining ecological balance, (c) Horticulture, in addition to supplementing the national foodgrid by providing nutritive food in the form of fresh and processed fruits and vegetables, can be an important source of foreign exchange earnings. Certain species of fruit trees, vegetables and medicinal plants, (e.g. mango, pomegranate, neem, jamun and guava) which can be grown on wastelands are an important source of agricultural exports. Thus afforestation supplemented with horticulture, floriculture, apiculture and food processing in urban areas can assure substantial amount of agricultural exports and afforestation supplemented with other related vocations such as bee keeping, livestock raising, charcoal making, carpentry, paper and pulp production can assure year round employment in rural areas. Efforts need to be made first, to carry out micro level surveys for systematic identification of the extent, location and nature of wastelands suitable for plantation work and second, define specific development projects (for horticulture, floriculture and apiculture etc.) that necessarily will entail site specific work. A processing plant with a marketing system should be designed along with this development plant to link it with a strong marketing system.

An aerial survey of some parts of India reveals requirement of further research to work out the economics of utilisation of (a) Chamba Ravines in Madhya Pradesh for production of tropical fruits (mainly mango), (b) tidal lands in Goa, Kerala and Karnataka for production of coconut and Banana. and (c) utilisation of community parks, river basins, reserved forests, district parks and educational institutions along the highways (on the lines of Tulip plantation in Netherlands) in urban areas for production of flowers and vegetables. Urban farming will reduce infrastructural cost as producer services are available in urban areas.

It is hoped that the present study would enable a focused thrust on enhancing the competitiveness of agricultural exports by adopting a suitable strategy of agricultural exports based on comparative advantage with major stress on improvement in productivity of agriculture, duly supported by efficiency in resource management, reduction in wastes and viable infrastructure in the agricultural sector.

Bibliography and Data Sources

Bibliography and Data Sources

Adleman, Irma. (1984), "Beyond Export led Growth". *Working paper No. 309, Department of Agriucltural and Resource Economics, University of California, Berkeley*. April, 1984.

Adndre, Sapir. (1983), "Factors Influencing India's Export Performance". Working Paper No.9, *Indian Council for Research and International Economic Relations, Lodi Estate,* New Delhi.

Agricultural and Processed Food Products Export Development Authority (1988), Conference Papers of *National Conference on Agro Exports* APEDA, Delhi.

Agricultural and Processed Food Products Export Development Authority. (1989) *"Indo-French Business Conference" Jan. 24, APEDA, Delhi.*

Ahluwalia, Isher Judge. (1991), *'Productivity and Growth in Indian Manufacturing'*, Oxford University Press, London.

Allen Maunder and Ulf Ren Berg. (1986), 'Agriculture in a Turbulent World Economy', *Proceedings of 19th International Conference of Agricultural Economists (1985),* Gower, England.

Asian Development Bank (1969), *'Asian Agricultural Survey'*, Economic and Development Research Centre, ADB Manila.

Asian Development Bank (1976), *'Asian Agricultural Survey, Rural Asia: Challenge and Opportunity'* Economics and Development Research Centre, ADB, Manila.

Asian Development Bank (1984), `*Regional Seminar on Agriculture'. Papers and Proceedings,* Economic and Development Research Centre, ADB, Manila.

Asian Development Bank (1986), `*Activities of OECD in 1985', Report by the Secretary General OECD,* Paris.

Asian Development Bank (1987), `*Towards Regional Cooperation in South Asia',* Economics and Develop:nent Research Centre, ADB, Manila.

Azam K.M., (1968), `*Planning and Economic Growth with Special Reference to Agriculture'*, Lahore, Pakistan.

Balassa Bela. (1964), `*Trade Prospects for Developing Countries'.* Economic Growth Centre, Yale University.

Balassa, Bela (1965), `Trade Liberalisation and "revealed" Comparative Advantage', *Manchester School of Economics and Social Studies,* Vol.33, No.2

Balassa Bela (1971), *"The Structure of Protection in Developing Countries"* , John Hopkins University Press, Baltimore.

Balassa, Bela. (1977a), A "Stages Approach" to Comparative Advantage' *World Bank Staff Working Paper No. 256,* World Bank, Washington.

Balassa, Bela. (1977b), "Revealed" Comparative Advantages Revisited: An Analysis of Relative Export Shares of the Industrial Countries', 1953-71, *Manchester School of Economic and Social Studies,* Vol.45, No.4.

Balassa Bela (1977c), `*Policy Reforms in Developing* Countries'. Pergamon, Oxford.

Balassa Bela (1978), `Export Incentives and Export Programmes in Developing Countries: A Comparative Analysis'. *Weltwirtschaft Liches Archiv (Kiel), Vol.114,* No.1.

Balassa, Bela. (1979), "The Changing Pattern of Comparative Advantage in Manufactured Goods". *Review of Economics and Statistics,* Vol.61, No.2

Balassa, Bela. (1981), *`Newly Industrializing countries in the World Economy',* Pergamon Press, New York.

Balassa, Bela (1982a), *"Development Strategies and Economic Performance: A Comparative Analysis of eleven Semi Industrial Economies",* World Bank, Washington.

Balassa, Bela. (1982b), "Newly Industrializing Developing Countries after the oil crisis". *World Bank Staff Working Paper No.437,* World Bank, Washington.

Balassa Bela (1982c), *`Development strategies in Semi Industrial Economies'* John Hopkins University Press, Baltimore.

Balassa Bela (1989a), *`Comparative Advantage, Trade Policy and Economic Development',* Wheatsheaf, Harvester.

Balassa, Bela (1989b), *`New Directions in the World Economy',* Macmillan, London.

Bale, Malcolm and Bowers, Alan. (1983), "A profile of World Bank Activity in Fruit and Vegetable Projects". *Country Policy Department Discussion Paper No.1983-18, The World Bank,* Washington.

Ballance, Robert, H. (1987), *`International Industry and Business',* Allen and Unwin, London.

Ballance, Robert, and Helmut Forstner, (1990), *Competing in a Global Economy: An Empirical Study of Specialisation and Trade in Manufactures* Unwin,Hyman, London.

Banerji, R. (1975), *'Exports of Manufactures from India*, J.C.B. MOHR Tubingen.

Barend A. de Vries (1982), "Export Promotion Policies", *World Bank Staff working Paper No. 313, Industrial Development and Finance Department,* World Bank, Washington.

Bergsten, C Fred and William R Clive (1982), *'Trade Policy in the 1980s'.* Institute for International Economics, Washington.

Bhagwati, Jagdish and Srinivasan, T.N. (1975), *'Foreign Trade Regimes and Economic Development : India',* National Bureau of Economic Research, University of New York.

Bharadwaj (Ranganath) (1962), *'Structural Basis of India's Foreign Trade'* University of Bombay, Bombay.

Bhattacharya, B., Ramesh Behl, L.D. Mago. (1994), "Agriculture Under the Uruguay Round - Possible Impact of Reduction in Domestic Support and Export Subsidies" in *Symposium on Uruguay Round on Agriculture, IIFT*, New Delhi.

Burki (Shahid Javed) and Laporte, Robert Ed. (1987), *Pakistan's Development Priorities, Choices* for the 1986, Lahore, Pakistan.

Chakravarty, S (1983) "Trade and Development ; Some Basic Issues", *International Social Science Journal* Vol. XXXV, No. 3.

Chakravarty, S. (1988), "Development Experience in South Asia", *Asian Development Review,* Vol.6, No.1.

Chaudhri, Rajiv (1985), "The Impact of changing Affluence on Diet and Demand Patterns for Agricultural commodities", *World Bank Staff Working paper no. 785*, World Bank, Washington.

Chaurasia R.A. (1988), *'Agro Industrial Development: A Strategy'*, Chugh Pub., Allahabad.

Dattatreyulu (1987), 'India's Agricultural Exports' *Foreign Trade Review,* Vol.22, pp.302-28.

Dattatreyulu (1988), 'India's Agricultural Exports' *Foreign Trade Review,* Vol.23.

Dattatreyulu (1994), "Uruguay Round : New Opportunities for Boosting Agro-Exports" *Symposium on Impact of Uruguay Round on Agro-Exports, IIFT,* New Delhi.

Deardorff, Alan, V. (1980), 'The General Validity of the Law of Comparative Advantage', *Journal of Political Economy,* Vol.88.

Desai Meghnad, Susanna Hoeber, Ruddph and A.Rudra, (1984), *Agrarian Power and Agriculture Productivity in South Asia'* Concept Pub., Delhi,

Dhar Biswajit and Mohanty S.K. (1994), "Prospects of Market Access for Developing Countries in Post Uruguay Round Agricultural Trade" in *Symposium on Uruguay Round Agreement on Agriculture, IIFT,* New Delhi.

Donges, J.B., (1976), "Industrialization Policies in 15 Semi Industrial Countries, *Weltwirtchaftliches Archive,* 12(4).

Economic and Research Foundation, (1986), *'Agricultural Exports Strategy: Problems and Prospects'*, Radiant Pub. Delhi.

FAO, (1978), *'FAO Studies in Agricultral Economics and Statistics 1952-77'*, FAO, Rome.

FAO, (1982), "Horticulture in The Mediterranean Area, Outlook for Production and Trade" *Commodity Bulletin Series* (42), FAO, Rome.

FAO, (1984), 'Promoting Agricultural Trade among Developing Countries', *Economic and Social Development Paper, (41), FAO,* Rome.

FAO, (1987), 'Agricultural Stabilization and Structural Adjustment Policies in Developing Countires'. *Economic and Social Development Paper No. 65, FAO,* Rome.

Federation of Indian Chamber of Commerce and Industry (1982), *"Report on Exports by a Delegation to Korea, Singapore, Taiwan and Hongkong"*, FICCI, New Delhi.

Final Act Embodying the Results of the Uruguay Round of Multilateral Trade Negotiations, GATT (1994), Marrakesh, 15 April.

GATT (1994), *Country Study: India* U.N.

Ghosh Pradip K, (1987), *'International Trade and Third World Development'* Cambridge University Press, U.S.A.

Government of India, *Economic Survey of India,* Various Issues.

Government of India, *Sixth Five Year Plan.*

Government of India, *Seventh Five Year Plan.*

Government of India, *Eighth Five Year Plan.*

Grossman and Helpman, E., (1990), `Comparative Advantage and Long run Growth' *American Economic Review* 80, pp.796-815.

Gruber, W.H., Dileep Mehta and Raymand Vernon, (1970), "The R & D factor in Industrial Trade and Investment of U.S. Industries" in Vernon Ed. `*Technology Factor in International Trade,* NBER, New York.

Gulati Ashok and Sharma Anil, (1994), "Agriculture under GATT: What it Holds for India?" in *Symposium on the Impact of Uruguay Round on Agro-Export, IIFT,* New Delhi.

Harkness, John and Kyle John F., (1974), "The Commodity Composition of International Trade": An Empirical Analysis *Oxford Economic Papers,* Vol.26.

Harkness, John and Kyle John F., (1979), "Factors Influencing U.S. Comparative Advantage" An anitomy in the theory of comparative advantage, *Journal of International Economics,* Vol.9.

Heckscher, Eli (1949), "The effects of Foreign Trade on the Distribution of Income". *Readings in the Theory of International Trade, Philadelphia,* 1949, p.285

Heller, Perter. S, (1976), "Factor Endowment change and Comparative Advantage: The case of Japan". *The Review of Economics and Statistics,* Vol.58.

Hicks, Norman, L and others, (1976), "A model of Trade and Growth for Developing World", *European Economic Review, Amsterdam,* Vol.7, April.

Hirsch Seev, *Location of Industry and International Competitiveness.*

Hirsch Seev (1974), "Capital or Technology? Confronting the Neo-factor Propositions and Neo-Technology Accounts of International Trade". *Weltwirtschaftliches Archiv,* Vol.110, No.4, pp. 535-63.

Hirsch, Seev, (1975), "The Product Cycle Model of International Trade: A Multi- country Cross Section Analysis". *Oxford Bulletin of Economics and Statistics.* Vol.37.

Hoda, A., (1987),*'Developing Countries in The International Trading System',* Concept Pub. Delhi.

Huffbauer, G.C.(1970), "The Impact of National Characteristics and Technology on the Commodity Composition of Trade in Manufactured Goods" in *Vernon, R. Ed., 'Technology Factor in International Trade', NBER, New York.*

Hussain Akmal (1988), *'Strategic Issues in Pakistan's Economic Policy',* Lahore, Pakistan.

Ichimura Ed.(1987), *'Development Strategies and Productivity Issues in Asia',* Tokyo, Asian Productivity Organization, Tokyo.

Iicher Carl and Witt Laurence (1980), *'The Place of Agriculture in Economic Development'.* McGraw Hill, New York.

Islam, A.B.M.M. Azizul (1984), "Application of the Principal of Comparative Advantage in the Choice of Agricultural Crops". *Economic Bulletin for Asia and the Pacific, 35(2):* 85-95.

James, William E, Naya Seiji and Meier, Gerald M (1987), *Asian Development: Economic Success and Policy Lessons,'*

International Centre for Economic Growth, University of Wisconson Press, USA.

Jhingron V.G., ed.2 (1982) *'Fish and Fisheries of India'*, Hindustan Publishing, Delhi 1982.

Johnson, D. Gale, (1973) *'World Agricultre in Disaray'*, Fontana, London.

Johnson, H.G. (1968), 'Alternative Maximization Policies for Developing Countries Exports of Primary Industry', *'Journal of Political Economy', May-June.*

Johnston, Bruce F, (1951), "Agricultural Productivity and Economic Development in Japan" *Journal of Political Economy,* Vol.59, Dec.

Joshi (1987), *'Institutional Aspects of Agricultural Development'*, Allied Publishers, Delhi

Kanesalingam, Vaid and Kemal, A.R. (1990), *'Trade Expansion in South Asia: Liberalization and mechanisms'*, Concept pub. Delhi.

Kaska, N. (1987) *'Technical Assistance Programme for PARC Horticulture'*, PARC, Islamabad.

Kenen, P. B. (1975), *'International Trade and Finance: Frontiers for Research'*, Cambridge University Press, England.

Kenen, P.B. (1980), *'Essays in International Economics'*, Princeton University Press, England.

Khan (1987), *'Fresh Perspectives on India and Pakistan: Essays on Economics Politics and Culture'*, Lahore, Pakistan.

Khan A.Rahman ed. (1987), "Studies on Strategy and Technique of Development Planning" *Readings in Development Economic Series, No.1, Pakistan Institute of Developing Economies,* Pakistan.

Kim, Chungsco (1983), "Evolution of Comparative Advantage: The Factor Proportions Theory in a Dynamic Prospective, *Keilerstudien* J.C.B.MOHR, Tubingen.

Kravis I.B.(1970), "Comment on Huffbauer, Gruber and Vernon", in Vernon ed. (1970), *`Technology Factor in International Trade'* National Bureau of Economic Research, New York.

Krishnendy (1990), *`Optimal Export Product Mix Under DR and Foreign Exchange Constraint: An Exercise for Indian Economy',* Centre for Economic Studies and Planning.

Krueger, A.O., (1978), "Alternative Trade Strategies and Employment in LDCs". *American Economic Review,* Vol.68, May.

Krueger, A.O., (1983), "Protectionist Exchange rate Distortions and Agricultural Trading Patterns". *Journal of Agricultural Economics,* 65(5), December.

Kruenger, A.O. (1984), "Comparative Advantage and Development Policy. 20 years later" in *`Essays on Economic Structure and Performance',* Academic Press, New York.

Krugman, Paul R. (1980), "Scale Economies, Product Differentiation and the Pattern of Trade", *American Economic Review*, Vol.70, pp. 950-59

Krugman, Paul R. (1981), "Interindustry Specialisation and the Gains from Trade", *Journal of Political Economy,* Vol.89, pp. 959-73

Krugman, Paul R. and Helpman, E. (1985), `*Market Structure and Foreign Trade*' Wheatsheaf, p. 261-63, Harvester.

Krugman, Paul R. (1990), `Increasing Returns and Economic Geography', *Journal of Political Economy*, pp. 483-99.

Leamer (1984), "*Sources of Comparative Advantage: Theory and Evidence*", MIT, Cambridge.

Leonteif, W (1953), "Factor Proportions and the structure of American Trade", *Review of Economics and Statistics* Vol. XXXVIII, Nov.

Leonteif, W (1956), "Domestic Production and Foreign Trade: The American Capital Position Re-examined", *Review of Economics and Statistics* Vol. XXXX.

Linder, Staffan, Bruenstan (1961), `*Essay on Trade and Transformation*' John Wiley and Sons, New York.

Maizels, (1984), "A Conceptual Framework for Analysis of Primary Commodity Markets", *World Development Review*, Jan., 12(1).

Malen Baum, Wilfred (1959), `India and China, Contrasts in Development Performance', *American Economic Review,* Vol.49.

Manmohan Singh (1964), `*India's Export Trends and the prospects for self sustained growth*', Oxford University Press, London.

Michael Michialey, (1976), "*Exports and Growth: An Empirical Investigation*", Mimeo, World Bank, Washington, D.C.

Ministry of Commerce, Government of India, (1992), `*Export-Impot Policy Ist April 1992-31st March 1997*', New Delhi.

Mohanty, P.K. (1994), Uruguay Round Agreement on Agriculture: An Overview" in *Symposium on the Impact of the Uruguay Round on Agro-Exports, IIFT,* New Delhi.

Monke, (1981), E - "Towards a U.S. Agricultural Export Policy for the 1980s, Discussion", *American Journal of Agricultural Economics* 63(5), Dec.

Mukherji, I.N. (1991), "Regional Trade, Investment and Economic Co-operation Among South, Asian Countries", *Centre for South Central, South East Asian and South West Pacific Studies*, JNU, New Delhi.

Mukherji, I.N. (1994), "The New Trading Order : Implications for India", *Paper presented at the Thirteenth European Conference on Modern South Asian Studies, Toulouse, 31st August - 3rd Sept.*

Mukherji, I.N. (1994) - "The New World Order: Access to Trade and Markets", Paper presented in the *Seminar on New World Trade Order: Prospects and Concerns, December 1994. Organized jointly by Konard Adenauer Foundation, Madras and Management Association, Centre for South and South East Asian Studies, Madras University.*

Mukherji Neela, (1988), `*BOP Adjustments of South Asian Economies*', Best Books, Calcutta.

Naya, Seiji, Urrutia, M. Mark, S., and Fuentes, A., (1989), *'Lessons in Development: A Comparative Study of Asia and Latin America'* International Centre for Economic Growth Publication, California.

Nayyar Deepak (1976), *'India's Exports and Export Policies in 1960s'.* Cambridge University Press, England.

Nayyar, Deepak and Sen, Abhijit (1994), "International Trade and the Agricultural Sector in India", *Economic and Political Weekly,* May 14.

News of the Uruguay Round of Multilateral Trade negotiations (1994): *The Final Act of the Uruguay Round: Press Summary, 5 April, 1994,* GATT.

Nurul Islam, (1968), Tariff Protection, Comparative Costs and Industrialization in Pakistan *-Pakistan Institute of Development Economic Research Report,* Islamabad.

Nurul Islam (1990), "Horticultural Exports of Developing Countries", *International Food Policy Research Institute, Research Report, 80 Washington, USA.* , April.

Nurul Islam (1991), "South Asia : Those who are left behind", *Asia Pacific Journal of Rural Development,* Vol. 1, No. 1.

Nurul Islam (1992), "Agricultural Trade Between Developing Countries : Patterns and Future Possibilities", *International Food Policy Research Institute, Washington, USA.*

Nyerere Julius, K.(1990), *'Challenge to the South',* Oxford University Press, London, 1990.

Pahariya, N.C. (1988), *'Indian Exports, Structure and Determinants'* Concept Pub. Delhi.

Panchmukhi, V.R., (1978), *'Trade Policies of India: A Quantitative Analysis'*, Concept Publication Company, Delhi.

Panchmukhi, V.R. (1983), 'Neglect of Economic Services', *Economic Times*, March 9.

Panchmukhi V.R. and Brahamananda (1987), *'Development Process of the Indian Economy'*. Himalaya Publishing House, Bombay.

Panchmukhi, V.R., Rao, L.L., Mukherji, I.N. (1990), *'Economic Co-operation in the SAARC Region, Potential, Constraints and Policies'*, Concept Pub. Delhi.

Pearson, R.S. and Meyer, R.K. (1974), "Comparative Advantage Among African Coffee Producers", *American Journal of Agricultural Economics,* 56 : pp.310-13.

Ramesh, C. Kumar (1987), Agricultural Surplus, Variable Land and Long-Run Growth in an Open Dual Economy, *Occasional Paper No. 15,* RIS, Lodi Estate, New Delhi.

Rana (1985), - *'Exports and Economic Growth in the Asian Region',* ADB, Manila.

Rana (1987), *'Changes in the Export Patterns of Developing Countries',* ADB, Manila.

Rao, V.L. (1982), *"Determinants of India's Exports"*, Ph.D. Thesis, University of Delhi.

Romeo M. Bautista (1990), "Development Strategies, Foreign Trade Regimes and Agricultural Incentives in Asia," *Journal of Asian Economics,* Vol. 1.

Ronald, W, Jones and Peter, B Kenen ed. (1984), *Handbook of International Economics,* Article by V. Deardorf on "Testing Trade Theories and Predicting Trade Flows".

Sachdev, S. (1991), "Agricultural Exports and Competitiveness : Inferences of Inida", *Foreign Trade Review,* April-Sep,.

Sachdev, S. (1993), `International Competitiveness and Agricultural Exports of India', *Indian Economic Review,* Vol. XXVIII, No.2, pp. 203-217.

Scholberg, Henry (1987), `*South Asia Biographical Dictionary Index',* 2V., IDC, Switzerland.

Sen, R.K., (1983), "An Analysis of Exports of A few Agricultural Products entering into Mass Consumption", *Indian Economic Journal,* Jan - March.

Sengupta, A.K., (1990), "Alternative Estimates of Factor Intensity and Net Saving (Cost) of Factors: Some Results for Indian Economy", *Indian Economic Review,* Vol.XXV, No.1.

Sengupta, S.K., (1986), "Export Prospects of Cereals, Fruits and Vegetables", *Report, Federation of Indian Chamber of Commerce and Industry,* Bombay.

Singer, Hans, (1979), "Policy Implications of the Lima Target", *Industry and Development,* No.3, pp. 17-32.

Singer, Hans Neelamber, Hatti and Rameshwar, Tandon, (1984), *'International Commodity Policy, New World Order Series Pt.1'*, Indus Publishing, New Delhi.

Singer, Hans Neelamber, Hatti and Rameshwar, Tandon, (1990), *'North South Trade in Manufactures, New World Order Series'*, Vol.7, Indus Publishing, New Delhi.

Singer, Hans Neelamber, Hatti and Rameshwar, Tandon, (1990), *'Trade Liberalization in the 1990s, New World Series'*, Vol.8, Indus Publishing, New Delhi.

Singh, D.R. (1983), "Proxy Effects of Market in India's Exports", *Indian Economic Journal,* Jan - March.

Srinivasan, T.N. (1974), 'India's Trade Regimes' in *Proceedings of Seminar on Trade Strategies for Economic Development: The Asian Experience',* ADB, Manila.

Staiger, Rober, W, Deardorff, Alan, Vand Stern, Robert.M (1987), "An Evaluation of Factor Endowments and Protection as Determinants of Japanese and American Foreign Trade", *Canadian Journal of Economics*, August, 20 (3), pp.449-64.

Stern & Leamer (1970), *'Quantitative International Economics'*, Aldine, Chicago

Stern, R. M. (1980), 'Policies for Trade and Development' in *Essays in International Economics*, Princeton University Press, England.

Stern, R.M. (1980), "Testing Trade Theories" in Kenen, P.B. (eds), *Essays in International Economics,* Princeton University Press, England.

Subramaniam A and Islam, N., (1989), "Agricultural Exports of Developing Countries : Estimates of Income and Price Elasticities of Demand and Supply", *Journal of Agricultural Economics Vol. 40,* No.2, May.

Subrahmanium, T.R., (1988), `*Establishment of National Institute for Horticulture in Pakistan*', Lahore, Pakistan.

Symposium on the Impact of the uruguay Round on Agro-Exports (1994), Indian Institute of Foreign Trade in Collaboration with Agricultural and Processed Food Products Exports Development Authority (APEDA), 11 August, New Delhi.

Tandon, Rameshwar and Hatti Neelamber, (1987) : *Exports and Development - The Indian Experienice.* Asia Publishing House, New Delhi.

The Economic Times (1994), "Limited Farm Trade Gains From GATT says FAO", 11 May.

The Economic Times (1994): Editorial, Opportunity, Not a Threat", 22 February.

Tiwari, R.S. (1986), `*India's Export Performance*', Deep and Deep Pub. Delhi.

Tiwari, R.S. and Bhushri, N.K. (1987), "World Trade Patterns: An Empirical Study of RCA Approach", *India Quarterly Journal of International Affairs.*

Tripathy, B.N., (1985), `*Exports and Economic Growth*', Delhi, Mittal Pub.

Tuong, H.D and Yeats, A (1988), "On factor proportions as a guide to the future composition of Developing countries exports". *Journal of Developmental Economics*, Dec., 7(4).

Tyler, W.G. (1981), "Growth and Expansion in Developing Countries: Some Empirical Evidence", *Journal of Developing Economies, Amsterdam,* Vol.9.

Ulrich Koester, Hartwig, Schafer, and Alberto Valdes, (198), "External Deamnd Constraints for Agricultral Exports", *International Food Policy Research Institute. Reprint No. 163, Washington, USA.*

UNIDO (1980), *'International Comparative Advantage in Manufacturing'*, New York, U.N.

UNIDO (1981), *'World Industry in the 1980s'*, Regular Issue of the Bienniel Industrial Development Survey, New York, U.N.

UNIDO (1985), *'Patterns of Trade and comparative Advantage'*, New York, U.N.

UNIDO (1985b), *'Industry in the 1980s, Structural Change and Interdependence'*, New York, U.N.

United Nations (1985), "Estimating and Forecasting of Trade Shares", *UN Development Paper No.5 (ESCAP).*

Verghese S.K. (1978), "Export Assistance Policy", *Economic and Political Weekly,* February.

Vernon, Raymond (1970), *'Technology Factor in International Trade'*, National Bureau of Economic Research, New York, 1970.

Wadhwa, C.E., *"Regional Economic Co-operation between SAARC and ASEAN and selected other countries of Asia Pacific Region"*, Centre for Policy Research, New Delhi.

Wolf, Martin (1982), *'India's Export'*, A World Bank Publication, Oxford University Press, England.

World Bank (1989), *'World Bank Country Study: India, Recent Developments and Medium Term Issues'*, World Bank, Washington.

Data Sources

1. *Year Book of International Trade Statistics,* Vol./I & II, Department of International Economic and Social Affairs, Statistical Office, United Nations.

2. *Monthly Statistics of Foreign Trade of India,* Department of Commercial Intelligence and Statistics, Vol. I, Calcutta.

3. *Production Year Book*, F.A.O. Statistical Series, No. 47, Food and Agricultural Organisation of the U.N., Rome.

4. *Trade Year Book,* F.A.O. Statistical Series, No. 35, Food and Agricultural Organisation of the U.N., Rome.

5. *Year Book of Labour Statistics*, International Labour Organisation, International Labour Office, Geneva.

6. *Key Indicators of Asian Developing Countries,* Asian Development Bank, Manila.

7. *Economic Survey of India*, Ministry of Finance, Govt. of India.

8. *International Financial Statistics,* International Monetory Fund, U.N.

9. *Foreign Trade Statistics of the Asia and the Far East,* United Nations.

10. *Hand Book of International Trade and Development statistics,* United Nations.

11. *Statistical Year Book,* UNESCO, U.N.

12. *Report on Currency and Finance,* Reserve Bank of India.